ARAWATA BILL

THE STORY OF LEGENDARY GOLD PROSPECTOR WILLIAM JAMES O'LEARY

IAN DOUGHERTY

EXISLE
PUBLISHING

ISBN 0-908988-15-X

Exisle Publishing Ltd
PO Box 60-490, Titirangi
Auckland 1230, New Zealand
email: mail@exisle.co.nz
http://www.exislepublishing.com

First published 1996 as part of Exisle Publishing's
New Zealand Lives series.

Cover design by Craig Humberstone.
Typeset in Adobe Garamond, Trajan and Helvetica.
Design and artwork by Streamline Creative Ltd, Auckland.
Maps by Jonette Surridge.

Printed by Colorcraft Ltd, Hong Kong.

To my parents.

CONTENTS

PREFACE

I first heard the name Arawata Bill one wet Easter Monday afternoon at the old Park Board hut which used to squat beside Lake Howden in the Fiordland National Park. I was a varsity student walking the Routeburn Track during the Easter break. Over an enamel mug of packet soup I listened to an old-timer reminisce, as old-timers do in back country huts on wet Fiordland afternoons. In the accidental sauna of stove drying sweaty woollen socks and bush shirts he told me of a character called Arawata Bill he'd met while working on the Milford Road during the Great Depression.

They were wonderful stories I hoped were true but had my doubts. Stories of a man who was born into poverty, lived on porridge, had a 40-year-old horse and climbed mountains in a three-piece suit and thigh gumboots in his seventies. They were stories of a man who wasted a lifetime in the remote valleys of North-West Otago and South Westland looking for lost ruby mines and hidden sovereigns and buried sea-boots full of gold. Stories of a man who was buried in a pauper's grave and somehow became a folk hero and a legend along the way.

The stories of Arawata Bill stayed with me and I added to the collection over the years. More elusive was material about the man behind the legend, William James O'Leary, and how he became that legendary figure in New Zealand's folklore and literature.

In the quest for payable gold about Bill O'Leary I am indebted to many people. Present or former residents of South Westland, Henry Buchanan, Allen Cron, Betty Eggeling, and Bill, Des and Kevin Nolan allowed me to mine their precious memories.

So too did present or former residents of the communities near the shores of Lake Wakatipu, Ernie and Harry Bryant, Albie Collins, Neil Cook, Jim Gilkison, Rosie Grant, Norrie Groves, Elfin McDonald, Andy Paulin, Peter Sharpe, Mark Shaw, Bob Thompson, Reta and Tommy Thomson, Lloyd Veint, Moira Webb, and Ailsa and Vic Yarker.

Former climbers Andy Anderson of Christchurch and Gordon Speden of Gore, Bill O'Leary's niece Stella Dalley of Wellington and Murray Gunn of Hollyford shared their invaluable recollections and Murray gave me access to his fire-rescued private papers. Information gathered by the late Peter Chandler and deposited in the Hocken Library in Dunedin provided a similar wealth of material.

For their assistance with information or advice I am also grateful to Brian Ahern of Wanaka, Helen Beattie of Fairlie, Julia Bradshaw of Albert Town, Dorothy Buchanan of Wellington, Alan Cooper of the Otago University Geology Department, Mavis Davidson of Leigh, Catherine Delahunty of Dunedin, Clark de Vries and Gary Hodson of Momona, George Griffiths of Otago Heritage Books, Jim Henderson of Auckland, Ann Irving of Waimatua, Brian Jackson of St Bathans, John Kelk of Invercargill, Graham Langton of Palmerston North, David McDonald of the Hocken Library, Sheila Natusch of Wellington, Alwyn Owen of Stokes Valley, Sister Duchesne of St Joseph's Cathedral Presbytery in Dunedin, Sister Florence of the Little Sisters of the Poor Sacred Heart Home in Dunedin, Ian Speden of Lower Hutt, John Thomson of the Victoria University English Department, and Gary Tricker of Greytown.

Ian Dougherty

Dunedin

2005

CHAPTER ONE

WETHERSTONS
& WAGGING

*T*here's slag-heaps of material from the early life of William James O'Leary that point to his becoming a gold prospector. Nothing to suggest he was of the stuff from which legends are made.

Gold lured his father, Joseph Timothy O'Leary, to the Antipodes. The Canadian-born gold-miner sailed first to the Victorian goldfields of Australia, then to New Zealand when news came of the discovery of gold at Gabriels Gully in Otago in May 1861. New fields followed, among them Wetherstons,[1] which took its name from two brothers who were not having any luck at Gabriels and chanced upon gold while pig-hunting in a nearby gully.[2] Timothy O'Leary, as he preferred to be known, arrived at Port Chalmers near Dunedin from Melbourne on the sailing ship *Empress Of The Seas* on 27 September 1861,[3] and carried his swag inland to Wetherstons, joining 5000 miners who had packed into the shanty town.[4]

In the biggest, busiest and most boisterous of the mining towns on the Tuapeka goldfields, Timothy O'Leary met Bridget Mary O'Connor. They were both of Irish Catholic stock.

HOCKEN LIBRARY

The gold-rush town of Wetherstons around the time Timothy O'Leary arrived in late 1861.

Timothy O'Leary was born on Canada's eastern seaboard about 1835 of Irish immigrant parents from County Cork. Mary O'Connor, as she liked to be called, was born in County Clare about 1839.[5]

With two of her cousins, Nancy and Bedilia O'Connor, the Irish domestic servant had fled the poverty and hardship of her birthplace and headed for the other side of the world as an assisted immigrant. The three colleens travelled first to Glasgow, where they boarded the *Pladda*. Three-and-a-half months out from the Clyde the migrant ship arrived at Port Chalmers on Boxing Day 1862. Crammed below decks were 368 passengers. They included 194 single female immigrants brought to the colony by the Provincial Government of Otago to try to redress the population imbalance in the wake of the influx of mainly unmarried male migrant miners.[6]

Nine months after Mary O'Connor stepped ashore at Port Chalmers, she and Timothy O'Leary took a 25-mile coach ride from Wetherstons to Milton.[7] There, on 25 August 1863, they were married, twice. The civil service was performed by the registrar of marriages for Tokomairiro, Alexander Ayson. The 28-year-old groom signed his own name. His 24-year-old bride scratched a cross on the page and the registrar signed on her behalf. They were then married by a Catholic priest, Father James Parle.[8]

Timothy and Mary O'Leary's marriage registration.

Timothy and Mary O'Leary set up their Wetherstons home near a large reservoir called the Phoenix Dam, which had been built to supply water to wash the miners' dirt. They had eight children, five boys then three girls. William James O'Leary, or Bill as he was called, was the second oldest, born on 28 October 1865.[9] A brother Joseph Timothy who was known as Joe had been born the previous year. After Bill came Daniel who was called Dan, Francis who was known as Frank, John who was nicknamed Jack, then Ellen, Mary Ann and finally Clara who was known as Clare, born on 19 September 1879.[10] She would have no memories of her father.

At first the Tuapeka goldfields lacked both a Catholic church and parish school to which their devout mother could send the older O'Leary children. Instead they attended the Wetherstons state school, which opened in 1868 in one of the 14 hotels the town had boasted in its brief heyday. By 1868 most of the pubs had closed with the departure of a largely fickle population to fresh gold discoveries elsewhere.[11]

It was another four years before Bill O'Leary and his brothers were able to clamber up onto a church cart and ride the two miles to St Patrick's church-school at Lawrence, which opened on St Patrick's Day 1872. Their teachers included John "J.J." Woods, the composer of the melody for the National Anthem, *God Defend New Zealand*. John Woods was headmaster of the parish school from 1874 to 1876[12] and possibly played a part in Bill O'Leary's life-long love of music.

Outside school hours there was plenty to keep the O'Leary children entertained. They could play in the once-crowded Broadway and Burke Street, or watch the smithy shoe horses, or muck about in the creek that came down the north side of the gully. There was adventure to be had exploring the diggings where their father had teamed up with sometimes six or seven other miners to build a dam and several miles of races to bring water to their claims.[13] When the boys were big enough to hold a rifle there were wild rabbits to be shot for sport and pot.[14]

An appreciation of music aside, Bill O'Leary left school with the barest of formal education. He liked reading, but grammar, spelling and punctuation were not his forté. This was at least partly due to the amount of time he spent away from school. He loved the bush and the hills, and frequently wagged, the classroom at school being no match for the classroom outside, despite the trouble in which it landed him.[15] After one period of absence, his first morning back at school saw him called to the front of the class, where the teacher stood with a rope complete with hangman's noose. The teacher placed the noose over the head of the terrified boy, who was convinced he was about to swing for it.[16]

Usually Bill O'Leary headed off alone. Once, as Alice McKenzie recalled of an incident he told her Martins Bay family:

> ...he ran away from school and home when he was about twelve with a companion a little older than himself. The other longed for the adventurous life of a bushranger and took his father's gun. The pair set up their headquarters in a cliff-side cave and sat round a fire in the evening to discuss their plans. Of course they were missed and the glare of their fire soon brought about their capture, and the would-be bushrangers were put to work.[17]

There was plenty of work to be done. There were domestic chores to be performed around the O'Leary house, where a new baby arrived every other year. There was a garden to be tended, wood to be chopped, hens to be fed, a cow to be milked. There were cattle to be looked after on the small area of freehold land his parents had named Phoenix Farm,[18] or on the commonage where the remaining miners were allowed to run cattle to supplement their mining activities.[19]

Bill O'Leary never talked much about his childhood, although he once told a workmate, West Coaster Jack Cron junior, that he had "a very hard upbringing".[20] For his mother the times were harder still. Not long after the youngest child Clare was born, Timothy O'Leary cleared out, leaving his

The O'Leary boys. Bill O'Leary (left) and three of his four brothers all dressed up for a formal portrait, about 1890.

wife and eight children behind. It had a profound effect on the family. The three girls in particular were bitter about being abandoned by their father. His name was never mentioned by them again.[21]

Timothy O'Leary headed for the West Coast, where he worked as a miner at Rimu, five miles south of Hokitika.[22] Mary O'Leary stayed in Wetherstons for a while, where she lived in increasing ill-health and poverty. As a token of independence and a guard against the vagrancy laws she kept a florin coin tied in the corner of her handkerchief.[23] Some time around 1886 or 1887 she took the younger children to live in Dunedin, where she rented houses in Arthur Street, Russell Street and Bay View Road.[24] By then her second-oldest son had begun the task of turning wandering into a way of life.

CHAPTER TWO

PASSES & PROSPECTING

*I*n the late 1870s Bill O'Leary left school with permission and for good. Away from Wetherstons he turned his hand to a variety of jobs throughout Otago, Southland and Westland, returning home at odd intervals to visit his mother and younger brothers and sisters.[1] The work reflected his Wetherstons childhood experiences of rabbits, cattle, bushcraft and gold.

According to Jack Cron junior, Bill O'Leary teamed up with his schooldays bushranger companion to shoot rabbits for a living for a few years, using a muzzle-loader gun. Jack Cron reckoned they became two of the best rifle shots in New Zealand and travelled to Australia to compete in a champion rifle shooting contest at the Sydney Exhibition, where they were placed second.[2] It's a story without corroboration.

Other jobs followed. As Bill O'Leary later reminisced to an *Otago Daily Times* newspaper reporter:

> I worked for a cattle rancher near St Bathans. There used to be some magnificent herds of cattle on the Maniototo Plains in those days. While there I used to go to all the mining centres around – Tinkers, Drybread, Welshman's Gully, and the others – and it was while there, too, that I went on my first trip to the Coast. A party of us went in from Queenstown and went right up to Greymouth.[3]

During this cattle droving trip he was captivated by the West Coast.[4] The love affair was to last a lifetime.

Drawn to the Coast, Bill O'Leary accepted a government contract to help keep open the Haast Pass bridle track, where slips and floods and fallen trees were common. It was important work. The track provided the vital communication link and stock route between Otago and the small communities of South Westland until the road was completed in 1960. He had sole responsibility for the section of track between the Haast Pass and Pringle Creek. He later described it as a lonely job. There was no one else for miles around.[5]

He also picked up casual employment wherever he could. In Northern Southland he worked as a labourer at Mossburn[6] and later as a packer for the musterers at Burwood and West Dome Stations.[7] He performed the same task for the musterers at Ayrburn Station in Central Otago.[8]

Increasingly this casual work was taken on to subsidise his abiding interest in his father's occupation. Whenever he had earned enough money to buy gear and provisions he would take off in search of gold. Unlike his father and the miners at Wetherstons who banded together to work a claim, he preferred the lot of the lone prospector.

Even when he was doing something else, Bill O'Leary's mind was as much on prospecting as on the job at hand. One of the musterers at Ayrburn Station, Dave Matheson, later wrote:

> ...every morning leaving camp he would tell us to bring in any cooking quartz. Of course we never bothered till getting near camp in the afternoon and we would pick up any sort of stone & next day he would grind them up & tell us there was no gold in any of them which we knew before we handed them to him.[9]

His early attempts to strike it rich took him to readily accessible areas such as Manorburn in Central Otago and over the Lindis Pass to Otematata in the Waitaki Valley of North Otago.[10]

In the Spring of 1897 the 31-year-old began to explore and fossick for gold in the vast expanse of remote, little-known and largely unmapped valleys of North-West Otago and South Westland. From the small settlement of Glenorchy at the head of Lake Wakatipu he carried his heavy pack up the Dart Valley and crossed over to the Arawhata Valley,[11] where he spent a few weeks prospecting.

The trip was considered newsworthy enough to rate a mention in the Queenstown newspaper, the *Lake Wakatip Mail*. It reported him as having said on his return to Glenorchy that:

> ...there is undoubtedly gold in the place, but so far he has only partially prospected the beaches as they are too deep for ordinary mining, but they appear to be no better than the beaches of the Upper Dart. However, he is going back shortly and intends to prospect some terraces of which he seems to have a very high opinion. Native game, such as kakapos, maori hens, etc., abound, but rabbits are not extraordinarily plentiful, and what there are are very tough. He reports the climate as being much milder than at the Head of the Lake and that it is only a days journey from the camps on the Upper Dart, from which it is only divided by a saddle. The country through which the upper waters of the Arawata run is said to be very rugged and difficult of access.[12]

Bill O'Leary's casual remark that it was only a day's journey from the Dart to the Arawhata from which it was only divided by a saddle, made light of his astonishing achievement. He had completed a return solo trans-alpine crossing from the Dart River up the Pass Burn and over the Barrier Range via a 6000 foot pass to the head of the Victor Creek tributary of the Joe River which flowed into the Arawhata. While the access was comparatively easy on the Dart side, on the Arawhata side he faced a steep and dangerous descent into the valley 3300 feet below and with no obvious route down the tussock ledges and vertical bluffs.

Just as astonishing as the feat was the fact that it had never been done before. Bill O'Leary was the first of few to cross the pass[13] which, following a

The head of the glacier-fed Victor Creek, with the steep slopes from O'Leary Pass to the right. Compare the advanced state of the glacier in this photograph taken by climber Albert Jackson in the 1930s, with a more recent photograph on page 114.

confused and confusing history, was named after him. The confusion began with Charlie 'Mr Explorer' Douglas, who later recorded:

> Learys Pass. I first found this pass when traversing the Arawata river in 1883. I got to the foot of it, but did not ascend it, and next year I sketched in the feature & roughly fixed its position from the top of Mt Ionia, and noticed that the ascent from Williamsons flat to the top of the pass was comparatively easy. ... I was not aware that any one had crossed this pass till a prospector named Leary came from the valley of the Dart into the Arawata sometime in 1897 via this route so I have called it by his name. As usual he gave no record of his journey so I can't tell what the route is like from the summit of the pass down to the Dart River.[14]

Charlie Douglas not only got the wrong name but got the wrong pass. The feature he named Learys Pass and said he sketched from the "Arawata upper flats",[15] is quite a distance to the north-east of the pass Bill O'Leary actually crossed.[16] It is probably the feature later named Waipara Saddle. The pass Bill O'Leary did cross, and which cannot be seen from Mount Ionia,[17] Charlie Douglas noted as Reilly's Pass.[18] O'Leary Pass, to the south-west of Mount Ian, was correctly named about 40 years after Charlie Douglas sketched the feature he later named Learys Pass in mistake for it.[19]

As if by way of apology and compensation for the mix-up, a further pass was subsequently named after Bill O'Leary. Climber Gordon Speden, who would get to know him well, perpetuated his name in a second O'Leary Pass to mark his pioneering of a route from the head of Lake McKerrow to the Kaipo River and Madagascar Beach on the West Coast.[20] This did not stop members of a later climbing party from claiming that to the best of their knowledge, theirs was the first recorded crossing of the pass. They did concede it was possible Bill O'Leary used the pass in the course of his prospecting, though they found no evidence of this.[21]

A campsite Bill O'Leary frequented while travelling through the Dart

Valley was also named after him. The picturesque and sheltered spot beneath Chinamans Bluff at the top end of Chinamans Flat became known as Arawata Bill's Campsite, after his later adopted nickname.[22] There's said to be a creek somewhere which bears his name too, but there's no record of where it might be.[23]

Bill O'Leary probably took the names in his stride, just as he took the physical features in his stride. There was work to be done, gold to be found.

In the following year, 1898, a prospecting expedition took him on his first trip to the remote coastal settlement of Martins Bay. There he met up with Daniel and Margaret McKenzie who with their five children, Helen, Dan, Alice, Malcolm and Hughie, were trying to extract a living from the enveloping bush. After a short stay he was off again up the coast to Jackson Bay.[24] Bill O'Leary became a frequent visitor to the area, prospecting for gold and helping the McKenzies keep open their cattle tracks, in exchange for gear and tucker.[25]

During these trips to Martins Bay and Jackson Bay, Bill O'Leary outlined what was to become his prospecting patch. Martins Bay took four days strenuous walking or riding from the shores of Lake Wakatipu via the cattle tracks of the Greenstone and Hollyford Valleys. If he was on foot, a partial shortcut was available up the Route Burn then over the Harris Saddle, but it tended to take just as long. If he was on foot and in luck, a rowing boat sometimes left at the head of Lake

I.G. SPEDEN

Daniel McKenzie used the back of his 1903 diary to keep tabs on how much Bill O'Leary owed for supplies and was owed for work.

McKerrow would take him the 12 miles down the lake, bypassing the notorious lakeside track known as the Demon Trail.

To reach Jackson Bay he would either travel from Martins Bay along the boulder-covered beaches and muddy bush trails to Big Bay, Browns Refuge and Barn Bay, or inland via Lake Alabaster and the mud, slush and swamps of the Pyke[26] and Cascade Valleys. A figure 8 was also available between Big Bay and the Pyke. Both routes then took him over the Martyr Saddle and down the Jackson River and lower reaches of the Arawhata. This too was a journey of several days.

The long way round went through country familiar to him from his days as cattle drover and trackman. It took him past Lake Hawea and Lake Wanaka, then up the bridle track alongside the Makarora River before crossing over the Haast Pass. On the other side flowed the turbulent Haast River, which had to be forded several times. There were also side streams to cross, which wavered between trickles and torrents depending on the time of year. On reaching the Tasman Sea he would continue down the coast over sand dunes and along beaches and bush tracks to Jackson Bay, fording the Okuru, Turnbull, Waiatoto[27] and Arawhata rivers along the way.

In between lay a huge expanse of rainforest sliced up by deep gorges formed by swift, flood-prone, quicksand-beached rivers flowing from snow-fed glaciers and snow-capped mountains rising above 8000 feet. It embraced the Barrier and Olivine Ranges and the highly mineralised and aptly named Red Hills Range standing bare among the surrounding bush-clad terrain. The area fascinated Bill O'Leary. During several decades he would acquire as good a knowledge of its nooks and crannies as any of his contemporaries.

CHAPTER THREE

PIGEONS
& PORRIDGE

*D*uring these early prospecting expeditions Bill O'Leary fashioned a way of life which enabled him to survive in a harsh terrain and extreme weather conditions for weeks and months at a time. The weather alone was enough to deter all but the hardiest of venturers. He faced blistering summer heat, intense rainfall, heavy snowfalls, tremendous gales and violent storms.

Sometimes, he later told a *Wide World Magazine* correspondent, the booming of the thunder "got a man down a bit", especially in the valleys when it echoed terrifyingly from peak to peak:

> I've seen great boulders brought down by the concussion of the thunder. And I've seen the lightning strike. I've seen rain that came down in sheets, turning the rivers into raging torrents as you looked. ... Yes, I could often tell what it was going to be like. When the "north-west" arch appeared in the sky, or hog's back clouds formed, you knew it was time to be making for shelter. But there were other times when the storms just seemed to come from nowhere – and they were always the worst. Still, it was sometimes grand to be caught out, and I shall never forget the time I saw a lightning flash splinter a great beech and fling the pieces all over the place. You don't sleep under trees in that country during thunderstorms.[1]

The shelter he would be making for varied. At the top end of the market were musterers' huts such as those established by Wakatipu runholders in the Greenstone Valley and those built by the McKenzies throughout the Hollyford. Away from the comparative luxury of the musterers' hut he would

SOUTHLAND DAILY NEWS, 15 NOV. 1947

Bill O'Leary outside his tent at the Pyke hut.

often build his own mai mai or makeshift shelters from whatever was handy. As he told a West Coast neighbour, Frank Heveldt, you weren't a bushman if you couldn't build up a bit of shelter.[2]

Sometimes he took a tent. They weren't flash. Arthur Hamilton knew something about tents. He and his brother Wilf ran a sports shop in Invercargill, as well as guiding parties through the Greenstone and Hollyford Valleys. Arthur Hamilton remembered seeing Bill O'Leary at Elfin Bay on the shores of Lake Wakatipu shortly after he emerged from the bush. He was drying an old tent he had bought second-hand in Invercargill several years earlier and was talking about buying a new one. According to Arthur Hamilton, "No one knew how he had existed under this tent, for no ordinary man would have been able to live in it."[3]

If he didn't have a tent he would be just as happy spending the night in a cave or beneath an overhanging rock.[4] Failing that, as he told Frank Heveldt, you could always get under a bank or poke in somewhere, even inside a hollow log.[5] Whatever the shelter, it was usually personalised. He would talk about "my cave" or "my mai mai".[6] Hut or cave, tent or mai mai, he never carried a sleeping bag. A blanket and his clothes were all he slept in.[7]

Those clothes were usually a work shirt worn under a heavyweight three-piece brown suit, complete with chain leading to a gold watch in his waistcoat pocket. Attached to the chain was a key for winding the watch through a hole in the back. Lose the key and the timepiece became mere ornament. On the front was a protective cover which could be flicked open with one hand to reveal the time.

On his head for protection from sun and rain he wore a hat. The rain didn't bother him and he had no wet-weather gear other than an old coat, but he liked to keep his head and shoulders dry.[8] The rest of him could easily be dried by the fire at night. While others preferred to hang their clothes in front of the fire, he would sit in soaking wet clothes facing away from the fire with clouds of steam coming off his back.[9]

On his feet he wore a pair of sturdy boots, or knee or thigh gumboots pulled up over his suit trousers. These were no ordinary gumboots. They would be taken brand-new to a cobbler who would add a leather sole and hobnails. They were not the most comfortable of footwear but were ideal for walking through wet bush and working in river and stream beds. He wore them winter and summer.

This working clobber was not much different from his Sunday best of three-piece suit, shirt and tie, dress hat, boots, watch chain and what would become a trademark flower in his jacket buttonhole.

Bill O'Leary undertook his prospecting expeditions either on foot with a pack on his back, or with a packhorse which he would ride or lead depending on the load and the terrain. Often he would pack in his supplies from inland Lake Wakatipu or the coastal settlement of Bruce Bay north of Haast.[10] Sometimes he would have them delivered by ship to Okuru south of Haast or to Big Bay or Martins Bay.

Even on foot he was able to carry in enough gear and food to last several

WIDE WORLD MAGAZINE. OCT. 1948

" Great boulders brought down by the con-
cussion of the thunder."

The making of the myth: how Wide World Magazine depicted Bill O'Leary caught in a thunderstorm.

months. This he did by making numerous trips over the same ground. He might walk for half a day with as much as he could carry, dump it, back-track and return the next day with another load. Once everything had been patiently relayed half a day ahead, he would do the same again in stages until he reached his destination.[11] Supplies and tools would also be stashed in caves or beneath overhanging rocks along the way.[12]

Descriptions of Bill O'Leary's size vary wildly and can only partly be explained by his getting smaller as he aged. Some remembered him as being a tall chap, all of six feet, and weighing up to 14 stone. Others recalled him as being a small bloke, about five foot six or five foot eight, and lean, wiry and all sinew.[13] Photographs and weight of numbers suggest he was closer to the latter than the former.

Differences over size aside, it was generally agreed that he was strong, athletic, energetic, hardy, unmatched for sheer rugged toughness, capable of carrying heavier weights than most people and very proud of his ability to do so. He would think nothing of setting out with a pack described as weighing anywhere from 60 to 120 pounds but he reckoned at 100 pounds.[14]

Once a pack was on his back he seldom took it off until he got where he was going. When he wanted a spell it was easier to lean against a rock or a tree.[15] One of his acquaintances from the head of Lake Wakatipu, Fergie Heffernan, even suggested that Bill O'Leary felt a bit out of place if he didn't have something stacked on his shoulders, which he was constantly flexing. These contrasted with his legs, which were likened to fence standards and were a source of wonder that under the weight of the loads he carried they didn't break off and stab him.[16]

A man could carry so much. A horse could carry so much more, provided the load was correctly placed. It was a skilled task. Good judgement was required to ensure the weight was evenly balanced in pack bags usually made from split sacks and placed on either side of the horse. The penalty for misjudgment was slipping loads, bucking horses and the threat of losing load and horse over the nearest bluff.

To secure the loads he would use flax, with knots tied so tightly that no one could undo them. To unload the horse he would have to cut the flax with a knife. Flax came in handy in other ways too. He used it to patch his saddles, described as so rough that no ordinary man could have ridden on them. Flax was also tied on the end of his horse's reins to lengthen the lead. If he was riding, he would sit behind the side sacks and balance an additional pack or two in front. Once he had his food and gear packed away, he never bothered much about looking after it or keeping it in any semblance of order.[17]

Bill O'Leary had a succession of horses. He treated them well. His most famous and most enduring horse was Dolly, a bay mare with a white blaze on her face. She was somewhat unkindly referred to by an Otago acquaintance, Tom Kennett, as "a dopey looking thing. What we call a West Coaster. Not renowned for their beauty".[18]

On his horse he would carry his mining equipment of pan, pick and improvised short-handled shovel, made by hacking most of the handle off a conventional shovel. Later he would take to sluicing with hose pipes, and sluice boxes made from ponga or tree ferns. He would also carry a small axe or slasher for track clearing, a tent, blanket, billy, tin plate and sometimes a rifle. There was no need for tent poles. A frame could easily be fashioned from the surrounding bush.[19]

There was one item which never went on an unridden horse and without which he never went far. The prized possession was a camp oven, a round cast iron pot with three legs, lid and a handle which folded down the side. This was carried on his back or by hand, too precious to trust to a horse which might fall and break the oven. It was easily done. No matter how careful the owner, age and regular heating and cooling of the oven would eventually result in its cracking, particularly round the leg joins from where liquids would escape and render the oven useless for boiling or stewing.[20]

Food consisted of the barest of essentials of oatmeal, flour, rice, split peas or beans, sugar, tea, salt, treacle or syrup and hard ships' biscuits. Oatmeal was the staple, from which he would make huge camp oven fulls of porridge,

The making of the legend: Bill and Dolly loaded with provisions for a three-month expedition to the Olivine River, about 1937.

with maybe a bit of black treacle. Too much to eat for breakfast but never wasted. It could always be given a stir and reheated or eaten cold that night.[21] The following day he might vary the fare with a billy full of rice, hot in the morning, cold at night. A large billy of tea freshly brewed in the morning would be brought to the boil at intervals throughout the day.[22]

Even on the move, a pot of porridge would not be wasted. Davie Gunn, who took over from the McKenzies running Hereford cattle in the Hollyford Valley, recalled:

Once I met him at the head of Lake McKerrow pulling his horse after him. He was pulling hard and had a lean on of about 45 degrees. In one hand he had a pot of porridge and in the other a rope. The porridge was what was left over from breakfast and every time he felt hungry he would have a dip at it.[23]

Davie Gunn also told of a further incident involving the movable feast that Bill O'Leary took on a route of his own from the Lower Pyke over the Humboldt Range and down to Lake Wakatipu:

On the way he used to spend one night in an ice cave on the tops. And all he ever carried with him on a trip like that was a billy full of porridge. No blankets, no other food. On the way back he would carry a 50-pound bag of flour, among other things.[24]

In a more sophisticated mood he might make damper by mixing flour, water and salt into a stiff dough and cooking it on the bottom of the camp oven, like girdle scones. The camp oven also turned out excellent bread. After a lot of kneading, the dough would be left for an hour or so in the three-legged oven placed over embers, some of which would be piled on the lid.

This basic diet would be supplemented by food he acquired along the

C. DE VRIES

Bill O'Leary taking care of domestic chores at one of the huts he frequented.

way. It was not the ornithologist but the gourmet who told the *Lake Wakatip Mail* on his return from the Arawhata Valley that rabbits were scarce and very tough but native game such as kakapo or native ground parrots, and "maori hens" or weka abound.[25]

Later on, a group of climbers described how as cook for the day at a hut he made them a camp oven full of a most delicious stew containing kereru or native pigeons, kaka or native parrots and whio or native blue mountain ducks.[26] At higher altitudes recklessly inquisitive kea or native alpine parrots were easily stoned for the stew pot. He would eventually confide to a relative that he had even tasted the New Zealand national bird, the kiwi, which he described as beautiful to eat.[27]

Native birds also featured in his version of a portable hangi or earth oven. Pigeons, together with small stones made red hot in a campfire, would be wrapped in layers of flour bags and sacking and placed in the horse's saddle bags. There the pigeons would remain until they had cooked, some distance down the track. If they were available, a few potatoes and onions would be added. He used the same cooking method for flapper ducks and trout.[28] The lakes and rivers were full of introduced trout, each weighing up to 16 pounds, and he often pulled up two large trout on the one line.[29]

In addition to trout, the native birds would be supplemented by eels and flounders. The eels were for medicinal purposes as well as nutrition. When he was feeling off-colour, eating eels was said to have cured him.[30] Not that sickness was a frequent problem. Taking a drink of water running off either naturally occurring arsenic or more likely arsenic left over from gold processing made him pretty crook one time. He also suffered from bronchitis, but a stint in the clear mountain air would take care of that.[31]

The birds and fish he caught were sometimes complemented by a man-made harvest. He was known to "poke a few spuds in along the beach" and if he happened to come along when they were grown, he'd have a ready supply of potatoes.[32] He was also said to have planted apple trees and gooseberry bushes throughout his patch. Members of a survey gang staying at the Sunny

Creek hut in the Hollyford Valley were intrigued when he asked them if they wanted to make an apple pie. Without a further word he disappeared. They thought he had gone for good, but two hours later he returned with a sack of apples from one of his trees.[33]

Bill O'Leary had a reputation for having a sweet tooth. He would pack in a 75-pound bag of sugar, as much sugar as flour, and the sugar would invariably run out first. Several spoonfuls went into each mug of tea. He was known to eat it by the handful.[34]

The sweet tooth was also to the fore in his passion for chocolate. This was illustrated by a couple of stories told about him when he was staying at the Barrier hut, near the junction of the Pyke and Barrier Rivers. On one occasion the skipper of a fishing vessel and a companion paid him a visit. Their plan to prime him with whisky and get him yarning fell through when the skipper handed over a huge slab of chocolate. Bill O'Leary made himself comfortable on a chair, ate the chocolate and then disappointed his two visitors by falling asleep.[35] Gordon Speden had a similar experience with his emergency rations. The climber poured Bill O'Leary a whisky and told him to help himself to two heavy chocolate bars on the bunk. He left the whisky, grabbed the chocolate and scoffed the lot.[36]

As well as chocolate, he had a fondness for cake. For many years afterwards he described the wonders of a cake made for him by Haast resident Norah Cron, adding that "It had raisins in it".[37] Women at the head of Lake Wakatipu would also send him on his way with a cake they had baked for him.[38] In similar vein, he eagerly took a slice of Christmas pudding offered by some mountaineers he came across. His eyes were said to have opened wide and longingly while he told them he hadn't tasted Christmas pudding for years.[39]

Such offerings were not universally appreciated. Whatever else his diet lacked, there was no shortage of roughage. So much so that more refined food came as a shock after living on such tough tucker, even for a man who had early on lost his teeth. One story told of him walking out through the Greenstone Valley to Elfin Bay and coming across a group who had travelled up Lake Wakatipu from Queenstown for a Sunday school picnic at Lake Rere:

Out of the bush staggered a heavily-laden and bearded man...just finishing another months-long one-man expedition into the Red Hills country. For all this time he had been living on half-cooked eels, campoven bread which turns to bricks after a few days, and other hard-chewing foods. One of the lady Sunday school teachers recognized "Arawata" and said, "Here, try a piece of our sponge cake." Arawata's hard gums took a mighty bite, and he exclaimed in surprise, "Just like fog."[40]

Later on he started carrying Sunshine Glaxo baby food. The storekeeper at Glenorchy had given him a tin by mistake and he had taken a liking to it.[41]

Most of Bill O'Leary's exploration was undertaken alone or with a horse. Initially he tried a dog for company, but it didn't work out. Feeding the dog was a problem. When he was near the coast he would let the dog loose among the lupins and driftwood in the sandhills to catch a rabbit or chase a seabird along the shore.[42] Away from the coast a dog became a liability, killing off potential sources of food such as weka.[43] The last straw was when he was forced to leave his dog behind while crossing O'Leary Pass. He never took a dog with him again.[44]

CHAPTER FOUR

FARMS & FERRYING

*F*rom the late 1890s Bill O'Leary had been making regular forays to the West Coast. About 1912 he moved to the Coast, where he stayed for nearly two decades, carrying out the usual mix of casual work and prospecting.

The 46-year-old searched for gold around Okarito Forks north of Franz Joseph,[1] and up the valley of the Landsborough River which flowed into the Haast. The quest took him to the head of the valley and into the tunnel of ice under the Landsborough Glacier.[2]

South Westland farmers called on his services for a good deal of casual work. He helped clear bush on farm blocks in a time-honoured procedure in which trees were felled, left to dry, burnt and then grass or turnip seed sown in the ashes. His experience with cattle was also put to use by the district's farmers, including Norah and Jack Cron senior and family who lived in one of the scattered farmhouses which made up the settlement of Haast.[3]

His methods were not always conventional. Jack Cron junior recalled how Bill O'Leary helped the family muster a mob of very wild cattle up a valley off the Haast River. Several previous attempts using dogs had resulted in the men being chased up trees. Bill O'Leary suggested an approach he'd seen work before. Every time the cattle chased the men up a tree, they should call off their dogs and fire a plug of gelignite just close enough to the cattle to scare them. With Bill O'Leary's help they spent all day following his advice. The cattle were unmoved. The party returned home disgusted. When they went in the next morning to have another go, the cattle were all lying out in the Haast riverbed ready for muster.[4]

Further down the coast at Martins Bay the two remaining McKenzies,

Malcolm and Hughie, also called on his expertise. According to Jack Cron junior the McKenzies' cattle were no more keen on being mustered than his father's mob in the Haast. Bill O'Leary put in three months among the wild cattle, which included big bullocks up to 10 years old. It was a waste of time. After they had finally driven the cattle into open country, the mob escaped back into their familiar beat of the Hollyford Valley at night, much to the disgust of Bill O'Leary who wanted an overnight guard kept on them.[5]

The McKenzie brothers later received his help in more dramatic circumstances. They had been bringing cattle up the Hollyford when Hughie had suddenly taken ill with a severe attack of pleurisy and pneumonia. When Bill O'Leary arrived at the Hidden Falls hut he found a delirious and fevered Hughie and his very worried brother. He immediately set off on foot over the Harris Saddle and down the Route Burn to the nearest telephone at Glenorchy to call for a doctor, who walked in the same way to treat the patient. Bill O'Leary meanwhile took horses the long way round via the Greenstone Valley for the doctor and his two companions to use on their way out.[6]

It was not a good time for the McKenzies. When Hughie was strong enough to ride out himself, he ended up in the Frankton Hospital near Queenstown. He had a septic leg from a deep burn caused when his brother Malcolm put heated stones around him to keep him warm while waiting for help at the hut. Already in the hospital were their other brother, Dan, who had broken one of his legs in a mining accident at Glenorchy, and their father Daniel who was dying.[7]

Bill O'Leary also cashed in on his backcountry knowledge and skills by acting as a guide. In 1912, for example, he and Jack Aitken from Paradise at the head of Lake Wakatipu took James Webster on an extended expedition. James Webster was the Presbyterian minister in Queenstown, and a keen tramper and climber. He paid Bill O'Leary and Jack Aitken to escort him by horse and foot from Paradise, over the Harris Saddle to Martins Bay, up the coast and over the Haast Pass to Lake Wanaka. Frustrated by delays, Bill

O'Leary only went as far as Okuru before returning south to Arawata, with the intention of making his way back to Paradise via the Arawata Saddle.[8]

As well as casual work, Bill O'Leary found more permanent employment on the Coast. He was taken on by the Westland County Council as a trackman on the Paringa bridle track which went north from Haast almost to Bruce Bay.[9] Some accounts also have him keeping clear the 37 miles of track between Martins Bay and Jackson Bay.[10] After a stint with the county council as trackman, he took another job with the county, this time as ferryman.

Bill O'Leary came in the wake of a fine tradition of ferrymen in South Westland, where drownings were common until they were appointed to nearly all of the rivers in those pre-bridge days. Most of the South Westland ferrymen were responsible for one river each. A German immigrant named Bill Burmeister had been the first on the Arawhata. Bill Hindley had manned the Waiatoto.[11] Bill O'Leary had the more complicated task of plying passengers across the lower reaches of both rivers.

His base was a hut on a sheltered verge near the mouth of the Waiatoto

Bill O'Leary (left) and neighbour Charlie Eggeling outside the Waiatoto ferryman's hut.

River where Hindley Creek joined the river's south bank. The site later disappeared when the Waiatoto changed course to the west. The hut with its iron roof and chimney and split ponga and iron walls was lined in the traditional New Zealand way with pages from those cherished magazines, the *New Zealand Free Lance* and the *New Zealand Weekly News*. They decorated, informed, and retained the warmth from the huge open fireplace where Bill O'Leary cooked his meals. Inside the hut was rough bush furniture made from carved wood and sacking, which in the case of the bed was covered in ponga leaves. The only luxury was a portable gramophone which he had bought during a visit to Dunedin, taken to Hawea by bus, and transported on horse back over the Haast Pass to Okuru, with the brand new gramophone balanced in front of him. Outside the hut were rambling roses and a prolific garden which produced vegetables and strawberries.[12]

WEEKLY PRESS, 14 OCT. 1926

Bill O'Leary outside his Waiatoto hut from where he operated his elaborate ferryman's beat.

From his hut the bright-eyed ferryman would row to the north side of the Waiatoto, ferry his passengers across, fetch his horse which he grazed on an island in the middle of the river, ride after the travellers eight miles south to the Arawhata, ferry them across, return the boat to the other side, ride back to the Waiatoto and return his horse to the island. The elaborate procedure was reversed for passengers coming the other way. If he was going to be away he would leave a boat on each bank of both rivers. After ferrying over their gear, one of the travellers would return the boat to the opposite bank before being

Hindley Creek merges with the Waiatoto River at the eroded site of Bill O'Leary's ferryman's hut.

picked up by the other boat. Solo travellers would have to tow one of the boats back across the river.[13]

Each of the 12-foot wooden ferry boats carried three or four passengers depending on how much gear they had. Sometimes several crossings were necessary. Gear and saddles went in the boat. Horses swam behind on a lead. They made for a slow trip. Full tide was the best time to cross, when the current was at its most sluggish. Even then the current was frequently so strong that the boat landed on the opposite bank some distance downstream and had to be towed back upstream to allow for the current when re-crossing. The worst time was after heavy rain or a warm westerly snow melt which could quickly turn the rivers into rages. At these times, Bill O'Leary, the prudent ferryman who couldn't swim, would sit tight.[14]

Potential customers were kept informed of his movements by a simple but effective device. At his hut door was a slate with the chalked notice "Have gone to" and then the words "Arawhata" and "Waitoto". This enabled him to indicate where he was by underlining the appropriate word rather than having to write out the name of the river each time.[15] The conventional method of

OTAGO WITNESS, 28 MAY 1929

Bill O'Leary (left), Agnes Cowan (centre), Eric James (right) and two members of a 1929 geological expedition to the Red Hills. The group was photographed at Okuru, where the Cowans provided accommodation for the expedition.

attracting the ferryman's attention was by "cooee" or whistle. Local cattleman and county council agent, Dinny Nolan, had a more effective method. "Nolan's Cooee", as it was called, involved setting off a stick of gelignite.[16]

The ferryman's job comprised short bursts of activity interspersed with long periods of waiting. The number of travellers calling on his ferrying skills might vary from three to four a week to one every other month. The locals preferred when possible to work the tides and cross the river mouths on horseback. There were busy periods though. In 1923, for example, a government survey party camped alongside his hut and made frequent use of his services.[17]

Other customers included the Makarora guide and packman Eric James, whose experience of horses had come from a London milk run. During two consecutive summers, in 1928 and 1929, he worked as a packman for geological expeditions to the Red Hills.[18] Eric James took the geologists over the Haast Pass and down the coast, where they met Bill O'Leary at Okuru and wearing his ferryman's hat at the Waiatoto and Arawhata Rivers. One of the party, Otago University geology lecturer Frank Turner, did not endear himself to the ferryman-prospector:

> He didn't like me very much, becuse he wanted to talk incessantly about all the gold there was in the Waiatoto and the Arawata, and I wasn't all that interested, and he thought I was a very poor sort of an individual to call myself a geologist.[19]

In return for his efforts Bill O'Leary was paid a salary variously reported to be £20, £52 and £75 a year. He was also entitled to charge passengers two shillings each, and six pence per horse.[20]

Travellers had a much better deal than that. Not only did he never charge, but he insisted on sharing mugs of tea and food with the often wet and weary travellers. His first response to a cooee was to wave his hat and disappear back into his hut, from where smoke would go up to signal the billy was on for a brew-up. There would be a cuppa too for those who forded rather than ferried the river. Sometimes the food would include flounders or silver sea-run trout which he caught with pipi or cockle bait at the mouth of the Waiatoto. As an act of generosity and professional pride, he gave anyone with wet feet a new pair of socks. If he was going to be away, he left the hut stocked with food, tea, sugar and condensed milk, and a note telling visitors to help themselves.[21]

Every so often, Bill O'Leary would ride from his Waiatoto hut to Okuru to get provisions, which were brought into the mouth of the Okuru River by scow several times a year. He would also call at Ad Cowan's house, which doubled as the post office, and collect the mail which it took the postie Charlie Smith several days to bring down the coast to Okuru by packhorse.[22]

When Bill O'Leary passed the Okuru school, the children were intrigued by the man in the peaked hat and old brown coat sitting straight in the saddle and holding upright in front of him a stick which rose and fell like a cavalryman's sword as he rode to a slow trot. The children were not above mimicking his peculiar riding style while overtaking him on their horses. They dubbed him "Hoopla Bill" on account of his perhaps nervous habit of starting off a conversation with a "hoopla" sound. He in turn was not above a bit of friendly teasing. Sometimes he would grab hold of one of the boys and pretend to carry him off, saying to the struggling child, "I could use a boy like you at Waitoto."[23]

When he called on their parents he never went empty-handed. He would take vegetables or strawberries from his garden, or for the children gifts of shells he'd picked up on the beach and pieces of mica he'd come across. In return, people would drop off at his hut food delicacies such as whitebait.[24]

The ferryman's reputation for kindness extended to clean living and clean language. He was courteous and well-mannered. He never cursed or swore. Frank Heveldt recalled how Bill O'Leary had arrived at the Heveldt hut at daylight one morning. He had spent most of the dark and wet night crawling on hands and knees while lost in heavy bush between the Waiatoto and Arawhata Rivers. Where others would have felt justified in uttering a stronger remark, Bill O'Leary reverted to his pet expression. "By Christmas I'm wet." He used the expression "By Christmas" to such as extent that it became another nickname.[25] Others remembered his strongest remark being "Oh golly" and "Oh by golly", usually uttered in conjunction with a hand clap.[26]

Not only did Bill O'Leary refrain from swearing, but he had few other social blemishes. He was never known to gamble, didn't smoke and, in the words of those who knew him, was not a man of the booze. The occasional whisky or glass of beer was downed more from politeness than fondness. Indeed after long periods of sobriety in the bush, one glass of home brew was said to be enough to knock him out.[27]

CHAPTER FIVE

THE THOUSAND ACRES

"Hoopla Bill" and "By Christmas" were transient titles. While he was on the Coast, Bill O'Leary was also given a nickname that stuck. It was the name Arawata Bill, gained at a time when the common spelling was minus the h of Arawhata, although it sometimes attracted the extra t of Arawatta.[1] He readily took to the name, preferring to be called Arawata as if it were his proper name.

The nickname was partly due to his work as ferryman on the Arawhata River and partly because of the time he spent prospecting up the Arawhata Valley. Whether on leave from the Westland County Council or as a fulltime prospector he would disappear up the valley for lengthy periods in the hunt for gold.

The headquarters for the mammoth undertaking were variously his ferryman's hut on the Waiatoto and a camp on a flat alongside the Arawhata. There he lived in a tent and tended an extensive vegetable garden where parsnips and currant bushes continued to grow, long after he had abandoned the site.[2] He also made use of a hut further up the valley about two miles above the lush pasture expanse of Callery Flat, where it was claimed he once snared 3000 rabbits to finance his prospecting expeditions.[3]

In the first stage of the operation he would assemble his gear and supplies. As well as basic food and camping equipment, he took picks, shovels, crowbars and an old grindstone. To these he added some rolls of condemned fire hosing from the Hokitika fire brigade and a nozzle said to be as big as a field gun, with which he intended to do some serious sluicing.[4]

He would then relay everything the 28 miles it was possible to ride or

A 1938 photograph of the Arawhata Valley where Bill O'Leary would disappear for months at a time prospecting for gold.

lead a horse up the Arawhata Valley to a gorge, at the foot of which he set up camp under a big flat overhanging rock. There he terraced the sloping dirt floor to create a series of flat areas at different levels on which to walk, sleep and store gear.

The dry natural rock shelter was later rediscovered by four climbers who had made their way down the gorge from the Olivine Ice Plateau in 1955. One of the party, Tom Carter, later recounted:

I picked up an old axe; its handle crumbled in my grasp. There was the remains of the bushman's bed. Two slabs, totara I believe, made the head and the foot. Grooves were cut across the grain, at close intervals along the upper surfaces of the slabs. Manuka poles had rested in these grooves and padded with moss would have made a comfortable bed. The poles were decayed and fallen to the ground. The crumbling ends of these

Two climbers on a 1948 expedition pause in front of a camp oven that Bill O'Leary left below Camp Oven Dome in the Five Fingers Range.

poles still rested in the grooves. Also in the shelter of the rock was an old benzine box and a length of canvas hose.[5]

The axe head, bed ends, box and hose later disappeared, leaving the terraced floor as the only lingering sign of Bill O'Leary's occupation.

This staging post was at the foot of the notorious Ten Hour Gorge, named by the explorer and Chief Surveyor for Westland, Gerhard Mueller, "because it takes that time to get through by crawling under or over the masses of detached rocks which range in size from a 'digger's hut' to a 'Courthouse'."[6] Deerstalkers without packs were reckoned to have gone through in little more than two hours, fully laden trampers more like five. An overloaded Bill O'Leary had good reason to name it the Fourteen Hour Gorge.[7]

Leaving his horse to graze on the flats below, he would spend many hours and days patiently lugging his gear through the narrow, noisy, gloomy gorge, where rapidly melting snow could turn the river into a death trap. In many places he was forced to construct rope and sapling ladders to get the gear around bluffs and up rock faces.[8]

The reward for the struggle though the gorge was to come out onto the 500 or so acres of McArthur Flat, then through a narrow section of valley to the open country of Williamson Flat, named after an earlier prospector, Andy Williamson. Bill O'Leary coined a name for it too. He called it the Thousand Acres.[9] The vast expanse was flanked by bush-covered hills, snow-capped peaks and the more accessible Camp Oven Dome, so-named because he left an old camp oven below the dome.[10] The rusty remains of a kettle and shovel which once belonged to him were also later found on Williamson Flat.[11]

When Bill O'Leary entered his Thousand Acres, to his right was the Joe River leading to O'Leary Pass. To his left was the Arawhata River and Arawhata Saddle, and beneath it the famous Arawhata Rock Bivouac. The massive overhanging rock was more commonly known by names which not only gave a clue to its size, but from where the visitor came. Wellingtonians for example called it the Wellington Railway Station. Gore climber Gordon Speden called it Fleming's Rock because it reminded him of his hometown's multi-storeyed Fleming's Creamoata Mill.[12]

This was familiar country for Bill O'Leary. He had first prospected here in 1897 after crossing O'Leary Pass from the Otago side. Then he stayed a few weeks. Now he stayed a few months, living in a rough hut made of split slabs of timber.[13]

He was away so long that his absences caused concern, to the extent that in May 1929 the 63-year-old was officially reported missing. Under headlines such as "Arawata Bill, Prospector Missing", several newspapers reported that he was overdue and a search party was setting out to find him.[14]

The stories quoted the controversial climber Samuel Turner, an eccentric English draper, who had received the news in Wellington by telegram:

'Arawata Bill', states Mr Turner, is an old man and he earned his name for his courageous explorations in the exceedingly rough country where the Arawata takes its rise. He usually sets out alone on his prospecting trips carrying a 70lb swag and pick and shovel in addition. For long periods Bill will disappear into the wilds and turn up all sound and well afterwards. Mr Turner, as an experienced alpinist, speaks with profound respect of the difficult character of the country into which Bill ventures alone; and of the character himself. Mr Turner says, "He has a most romantic and interesting personality and everyone likes and respects him. Probably no explorer in this country has ever been through the hardships suffered by Arawata Bill, who is a well-known personage in South Westland. I should think that if a search party is preparing to look for him that there is grave anxiety as to his whereabouts and condition, for strong as he is, he is well advanced in years."[15]

In the same issue as the weekly Dunedin newspaper, the *Otago Witness*, reported him missing, it was able to tell its readers that a further telegram had been received in Wellington that "the veteran prospector and explorer of the South Westland mountain vastness" had turned up at Okuru. It added that he had been missing for many days and when he had set out it was known that he carried provisions for only a few days and was not in the best of health at the time.[16]

This newspaper publicity built on the regional fame he had earlier received. Alongside the story of him turning up, the *Otago Witness* reproduced a photograph of him it had first published the previous year.

The 1928 Otago Witness *photograph that contributed to the regional fame of "Arawatta Bill".*

The caption appended to the August 1928 *Witness* illustration had read:

> A veteran of the mountains. Arawatta Bill, an old West Coast miner, aged 73 years, photographed at Lake Hawea before setting out on his 100-mile trek over the Haast Pass, carrying a pack weighing well over half a hundred weight.[17]

As well as giving an added edge to his exploits by adding more than a decade to his age, the caption illustrates the extent to which his nickname had caught on, referring to him only as "Arawatta Bill".

The nickname had long since been ingrained in South Westland, where they relate a more personalised and detailed account of possibly the same incident. According to the local version, Bill O'Leary was overdue and the police asked Dinny Nolan's brother Jim, to try to get through the Ten Hour Gorge and see what had happened. Jim Nolan farmed sheep at Callery Flat in the lower Arawhata Valley but had never been through the gorge.[18]

When he finally came out onto the deeply snow-covered Williamson Flat, in the distance he spotted the lone prospector working away with a shovel. Jim Nolan thought he would creep along the edge of the beech forest and see what Bill O'Leary was up to. When he was about 100 yards away, he lit his pipe and was startled to see the distant figure suddenly straighten up, turn around and roar, "Come out of there. I know you're there. Come out and show yourself." With senses described as keen as a deer, the non-smoker had smelt the intruder's tobacco smoke in the clear mountain air.[19]

Bill O'Leary was said to be in a desperate state, half-starved and thin as a rake, his food gone, his rifle next to useless. Jim Nolan took his own rifle, shot one or two of the rabbits which were flopping around in the snow, and started stewing them in a kerosene tin over a big fire. The prospector was so hungry that he couldn't wait for the rabbits to cook. He was dipping the fat and scum off the top of the tin and supping it down scalding hot, saying, "Soup, lovely soup".[20]

Bill O'Leary's rifle was a legend in its own right. Some reckoned it was so hopeless that it could shoot around corners.[21] When Dinny Nolan had the rifle properly sighted, it took the unimpressed owner some time to break the habit of having to aim several feet to one side of his target to hit anything.[22] Filling the barrel with boiled mutton fat as a form of weather proofing probably didn't help either.[23]

Not long after the 1929 lost and found incident, Bill O'Leary left South Westland for his former haunts of North-West Otago. Prior to his departure from the district, his neighbour, friend and county council boss, Dinny Nolan, presented him with a poem. The Okuru bard used to make these up during the long hours cattlemen spend on horseback. The moving tribute to the respect and affection Bill O'Leary had earned among travellers and locals alike must have brought a lump to his throat. It read:

Bill has gone and left us,
Now a vacant hut you'll find,
But kindly recollections
With us all he's left behind;
He was a genial spirit,
Straight and honest, true as steel,
And the loss our district has sustained,
Everyone must feel.

Bill has gone and left us,
But his records plainly tell,
The Council never had a man
To do the job so well;
He was a good and faithful servant,
Both competent and skilled,
And our task is one of magnitude.
To find another Bill.

Bill has gone and left us,
But his successors find a task
To fill the post so ably
As the man we had there last;
And I cannot help remarking,
But the facts are very plain,
That South Westland's loss
Is North Otago's gain.

Bill has gone and left us,
Now the traveller he won't find
That cup of tea, that blazing hearth,
That welcome always kind;
But he has left his kettle there
Behind him, and provender on the shelf,
So when the weary traveller comes along,
He can help himself.

Bill has gone and left us,
But there're not many gorges black
That he hasn't blazed a trail,
Or exploration track.

And the hardships that he has endured,
No one knows but he,
No one knows those gorges
Of such inaccessibility;
And soon he will be off again –
He's got the fever still,
And the solitude will be disturbed
By Arawata Bill.

Bill has gone and left us,
And we hope he will make his "pile"
He seeks the precious metal
In back country yet awhile;
And we'll pray that he's successful,
For if everyone had their due,
Old Bill would find a nugget,
Fully two and three feet through.

I often think and ponder,
You'll pardon me, I ken,
But will our rising generation
Produce such Grand Old Men.[24]

Official records of the grand old man's term with the Westland County Council were lost in a fire, but the county clerk, Edward Stoop, later reinforced Dinny Nolan's tribute. He wrote, "I can truly say that his rating was 100% with all and he was held in the highest esteem by all those people that travelled the far south many years ago."[25]

CHAPTER SIX

THE FRENCHMAN'S GOLD

*B*ill O'Leary's search for gold was not confined to natural deposits. When he left South Westland it was without a legendary boot full of buried gold, sometimes called the Frenchman's Gold, and without a horde of gold coins, known as McArthur's Sovereigns.

The origins of the boot full of gold vary from storyteller to storyteller. Most share common elements of stolen gold, mutinies, shipwrecks, loot buried in a sea-boot inside a cave, one survivor passing on a location map just before he dies, a slip covering up the cave entrance and unsuccessful searches for the treasure.

Some of the stories have a French connection and varying degrees of detail. Generally they involve a mutiny on board a French barque which was carrying either gold coin or gold ingots. One story has this taking place in the early half of the 19th century, one around 1840, while another emphatically states that the year was 1830. The location is generally agreed to have been somewhere near the mouth of the Cascade River.[1]

According to some accounts, there were five survivors of the mutiny. They made their way ashore with a considerable amount of gold coin, which they hid inside a sea-boot and buried in a cave.[2] A more elaborate account has six of the rebel sailors surviving after the ship was driven ashore by a violent storm. The remains of the wooden ship were said to have been visible for many years, half-buried in the sand dunes. The men salvaged one of the ship's lifeboats, some provisions and enough gold ingots to fill two four-gallon kerosene tins and a sea-boot.[3]

There are a couple of versions of what happened next. In one they set off up the Cascade and Martyr Rivers to Laschelles Creek, said to have been named after them. When the creek became too shallow to take the boat any further, they buried the gold in the bush near the stream. To mark the spot they worked a cross-cut saw and drove a pick into a beech or birch tree.[4]

Many years later, the saw and pick were said to have been sighted by an early Cascade Valley settler named Frazer, while he was looking for cattle in the vicinity of Laschelles Creek. By then, the tree had grown around the tools and only the ends of the saw and pick were visible. When a subsequent search of the creek bank failed to find the tree, it was concluded that the creek had gouged the bank and toppled the tree into the creek bed.[5]

The second version has the six mutineers travelling about 20 miles up the Cascade River, way past the confluence with the Martyr River. When the water became too shallow they hid the gold in a cave near McKay Creek, before returning to the Cascade Valley.[6]

Having planted the gold, one account has two of the men drowning when they were washed out to sea while trying to cross the Waiatoto River in a raft. All but one of the rest made their way up the coast to Nelson and across Cook Strait to Wellington. In desperation they reported to the French authorities and were arrested, charged with mutiny and shipped back to France to stand trial.[7] Another account has all but one of the mutineers making their way up the coast to Greymouth, only to be handed over to the captain of a French warship, who hanged them on the spot.[8]

There's general agreement that one of the men fell ill and took shelter in a small Maori settlement, thought to be Bruce Bay.[9] He is said to have settled in Westport, where he lived in poverty, never able to save enough money for the charter of a vessel to pick up the hidden gold. Before he died, he handed his son a map showing the location of the gold. Ridicule rather than poverty prevented the son from fitting out an expedition from among the West Coast miners to whom he went with his story. It was left to the people of Dunedin to outfit a party, which sailed to Jackson Bay but was unable to locate the treasure. The rusty remnants of their strenuous but futile efforts were said to have been seen scattered about the site long after they abandoned the search.[10]

A variation has him moving north to Nelson, where as an old man he was cared for by a local family. Before he died he gave them a plan of the location of the cave in which the gold was hidden. A party from Nelson travelled overland to the Cascade from Otago. They could not find the cave and concluded that it had been covered by a slip. Bad weather and a shortage of food defeated their efforts to tunnel into the slip. They packed up and left, never to return.[11]

Some of the stories about the so-called Frenchman's Gold have nothing to do with a Frenchman, the French or France. Generally they involve stolen gold from Australia.

According to one, more than £30,000 worth of gold was stolen from the Australian goldfields around the 1870s. The thieves sailed across the Tasman Sea to the West Coast of the South Island. Somewhere near the mouth of the Arawhata River two men went ashore with the gold packed in a sea-boot. Their instructions were to bury it in a safe place and return to the vessel. Instead, the two decided to double-cross their comrades. After hiding the gold, they hid themselves in the bush. When they failed to return, the rest of the crew sailed away in despair. One of the men died from hardship before they could reach civilization. The other finally stumbled into a settlement but died before he could return to recover the loot.[12]

Other versions have three or four Australians robbing a bank or a convoy of about £10,000 worth of gold and hiding their booty before the police caught up with them. After serving a few months of a long sentence, they escaped, recovered the stolen gold and set out in an open boat for New Zealand. They landed somewhere around the mouth of the Cascade River, where they hid the gold in a sea-boot. When the going got tough they split up, each heading in the direction he thought would take him to civilisation. Only one survived, to reach Otago somewhere around Lake Wanaka.[13]

One twist has him dying before he could give an adequate description of the hiding place. Another has him being befriended by a family named

McLean. Before he died the following year he gave them a map showing where the boot full of gold was hidden. They organised two unsuccessful expeditions and concluded that the gold might have been buried by a huge and recent looking slip.[14]

Yet another variation has three men about 1890 stealing about £40,000 worth of gold dust or gold bars either from Sydney or from a mail-steamer in Melbourne. They rifled the strongroom before the ship sailed, removed the gold from three boxes, replaced the gold with lead and the loss was not discovered until the vessel reached London. The thieves then stole a cutter and sailed across the Tasman, landing at the mouth of the Cascade River.[15]

One of the men was either drowned when they were coming ashore the first night, or when the ship was battered to pieces in a storm while the other two were ashore burying the gold in a sea-boot in a carefully marked cave. The two survivors then separated. One went south and was never seen again. The other went north and was found and taken by ship to Dunedin, where he only lasted a day or two. Before he died he told his story to an old miner, who was either given or made a chart of the location of the cave.[16]

The miner organised a search party, which was joined by two West Coast explorers, Harold Hubber and Robert Paulin. Hubber was not confident of finding the gold, but Paulin assured them it was there. He had verified the story in Melbourne and signs of the wreck had been found exactly where it was supposed the ship had come to grief. As in many of the stories, a huge slip had come down at the spot where the cave was supposed to be and the treasure seekers did not dig far enough into the thousands of tons of soil and boulders to recover the gold. Hubber is said to have made a second attempt to locate the gold in 1921, but as fast as the old slip could be cleared, more rocks and soil slipped down.[17]

Some of the stories combine Australian gold and a French connection. One for example tells of the wrecking of a ship carrying stolen gold from Australia, and only one man, a Frenchman, managing to get ashore to hide the gold in a sea-boot in a cave. Before he died he gave a location map to a man from Hyde in Otago, a Mr McKay, who had saved his life. McKay led three expeditions to look for the gold, but again the cave had been covered by a slip, into which they tunnelled a side shaft but without success.[18] Other tales given currency include the gold coming from a pirate ship wrecked at Browns Refuge south of Cascade.[19]

Of all the stories in circulation, Bill O'Leary is said to have favoured one or more of the Australian versions, although at times he was somewhat hazy as to the details.[20] The story he told Davie Gunn was the one about the four Australian convoy robbers carrying the gold inland in a sea-boot and the sole survivor giving a map to the McLean family. Bill O'Leary thought he knew the route the robbers followed from the coast and reckoned to have found old axe marks or scars on two big trees. These would have been put there by the robbers to mark the hiding place. His explanation was that they would have tried to carry the gold inland until they realised the nature of the ranges ahead.[21]

Another vague account has him searching for a grove of large fern trees which he believed were "of South Sea Island origin and an important landmark in connection with the supposed hidden gold".[22] A variation has the robbers planting a pine tree which they had brought with them to mark the spot.[23]

Hard evidence for the existence of the boot full of gold is as elusive as the gold itself and for every believer there have been a dozen doubters. Some reckoned it was another of those embellished West Coast yarns. Some wrote it off as a huge hoax.

Bill O'Leary was one of the believers and in that sense it didn't really matter whether the gold existed. His firm belief was more important than any unknown truth. He was convinced the cache was there to be found, searched for it, even claimed to have been "within measurable distance of it". Some suggested it became an obsession. He was quoted as saying, "I've just got to lay my hand on it." Another time he was less engrossed, conceding that he certainly looked for the gold, but it never kept him there.[24]

A second story of buried treasure which took Bill O'Leary's fancy during his days on the West Coast has more consistency and greater credibility. It's the story of McArthur's Sovereigns.

The central character is a man called Duncan McArthur, who arrived at Jackson Bay in 1875 to farm cattle and sheep at the settlement of Arawhata. McArthur didn't trust the banks and insisted on being paid in gold sovereigns for the butter and mutton he sold to the locals. As he grew older the bachelor invited a mate, Jimmy Thomson, to help run the farm. He told Thomson he had saved 1000 sovereigns which he would leave him, along with the property, when he died. Until then, Thomson would pay for food, supplies, everything.[25]

Eventually, McArthur became senile. One night Thomson arrived home to be told by McArthur that he had the rooster in the pot. He did, feathers and all. Just before he died, he tried to tell his mate the location of the sovereigns but Thomson couldn't understand a word the dying man was mumbling.[26]

The local gossip was that the sovereigns were buried near what became known as McArthur's chair. Apparently, every fine Sunday for years McArthur would climb a hill near his house to a seat fashioned out of the roots of a large distorted beech tree, from where he could look out over the sea, and read his Bible.[27]

McArthur's Sovereigns, like the Frenchman's Gold, were never found by Bill O'Leary, or anyone else. Some say he did find a third treasure, a mine full of rubies which was shrouded in mystery and suggestions of murder.

CHAPTER SEVEN

THE LOST RUBY MINE

*T*he story of the lost ruby mine first came to public attention the year Bill O'Leary made his initial crossing of O'Leary Pass. On 4 February 1897 readers of the *Otago Witness* were presented with an amazing article headed "A Secret Of The Olivine Range" and written by someone with the initials R.S.[1] The writer explained that he had persuaded a friend called Walter Murray (not his real name; none of the names were) "to write the account of his valuable discovery and experiences connected therewith." R.S. then produced what he called Walter Murray's Narrative.

In the lengthy and minutely detailed manuscript, Murray stated that he was a native of Norwich in England, where he trained to be a lawyer. It was not a calling to his taste. Instead he sailed for New Zealand in 1889, stayed a short time, then went to Australia where he worked at a mine battery at Ballarat and attended the school of mines. There he befriended one David McLean, with whom he worked in several tunnelling contracts and did some prospecting in the New England district. The two then separated, McLean going to the Queensland opal fields while Murray went to work at a battery at Hillgrove.

Two years later Murray returned to New Zealand with two other men. Prompted by a newspaper report of a deposit of nickel ore on the West Coast, the three tried their luck at Big Bay and Barn Bay. From there, Murray and one of his companions went up the Cascade River. The companion returned to Barn Bay. Murray decided to make his way across to Lake Wakatipu.

During the trip, Murray accidentally discovered dark red rubies enclosed in hard green rock in a gap in a cliff-face east of Lake Wilmot. Using the hammer end of his pick, Murray recovered "two large well-coloured stones about the size of marbles, seven others none less than five-sixteenths of an inch in diameter, and several smaller stones the size of peas."

Murray completed the journey over the Barrier Range to Lake Wakatipu and on to Dunedin, from where he wrote to his friend McLean in Australia. He enclosed two of the rubies, with a request that McLean have them valued and then join him in New Zealand for a return trip to the site of the discovery. While Murray waited for McLean, he stayed with his friend R.S. near Oamaru in North Otago. McLean arrived from Australia with good news. An expert in Melbourne had pronounced the two samples genuine stones and of a greater value than Murray had expected.

On the voyage to New Zealand, McLean had met a man called Cray. The two friends decided to take him on as a partner, even though Murray doubted his trustworthiness. The three travelled to Glenorchy, packed their gear up the Dart on horseback and relayed it by foot over the Barrier Range to the location of the find.

In their first week of mining they collected "more rubies than would fill a half pint measure" and a few good sapphires. Cray then had a serious accident. He fell about 25 feet from a ladder propped up against the huge wooden structure the three had built from trees to create a working platform on the side of the cliff. Cray suffered leg and head injuries, which resulted in severe pain and fever. He became delirious and raved that he was going to kill the other two and keep the rubies for himself.

Worse was to follow. When Cray recovered from his injuries he apparently tried to poison the other two. Murray returned unexpectedly to the tent where Cray was cooking dinner and found Cray emptying the contents of a small paper packet into the porridge pot. Cray claimed it was salt, but when dinnertime came, Cray said he had burnt the porridge and had to throw it away.

Soon after, Murray was lying down on a rug outside the tent having a nap when, in Murray's words, the brewing storm burst:

How long I slept I do not know – perhaps an hour or more – when I woke suddenly out of a troubled dream with the report of a revolver ringing in my ears and a heavy body falling across my chest, knocking the wind out of me for a few seconds. When I recovered my senses completely, to my dismay I found McLean, a smoking revolver in his hand, hauling the inanimate body of Cray from on top of me. McLean's face was white as chalk as the two of us stood looking at what lay at our feet. Cray lay there an inert mass – dead – with a bullet hole in his left temple, from which blood was trickling. [McLean explained] "As I came round that bush there," pointing to a spot about 12 yards away, "I saw Cray in the act of striking you with the axe. He was standing over you as you lay, and the blade shone as he swung it up to strike. I had no time to think over things; I tried to call out as I cocked the revolver, but my voice seemed to stick in my throat, and I fired. The whole thing happened in a second or two."

They buried Cray in the scrub under a crude cross and stone cairn and returned to Glenorchy with "wealth enough to make us both comfortably well off for life". They told the man from whom they had hired the pack-horses that Cray had decided to remain in the ranges for a time.

Murray ended his manuscript by explaining that he and McLean had decided to wait until they were safely away from New Zealand before telling the authorities what had happened. Otherwise there would have been an enquiry which would have detained them until a search party was able to verify their story the following summer. To help with the eventual search, Murray included a rough sketch map of the track he had followed from the Cascade and Pyke Rivers over the Barrier Range to the Dart River and Lake Wakatipu.

R.S. appended to the manuscript a letter he had received from Murray from the Arundel Hotel in London, dated 6 August 1896. In the letter Murray gave R.S. permission to make public the manuscript, which Murray had left with R.S. with a promise not to reveal its contents until Murray gave him the go-ahead. Murray explained that McLean was in Amsterdam negotiating the sale of some of the rubies, and that Murray had also "sold a goodly number of the stones, and the proceeds exceed our wildest hopes". He added that they were keeping some of the rubies for later disposal, although McLean was having three of the larger ones cut and set for R.S.'s wife.

It's a credible story. If someone made it up, they went to an extraordinary amount of trouble for no apparent reason. The detail alone suggests authenticity, even if some of the details stretch the bounds of credibility. Maybe, for example, the death of Cray didn't quite happen the way Murray described. How fortunate Murray and McLean were that Murray walked into the tent just as Cray was pouring the poison into the porridge. How fortunate they were that McLean arrived just as Cray was in the act of swinging the axe over Murray's napping body. How fortunate they were that McLean was carrying his loaded revolver and was such a good or lucky marksman that he hit Cray in the temple with one shot from about 12 yards.

As with the Frenchman's Gold, different versions have subsequently been told of these events, which would have taken place in the early 1890s. Some bear some resemblance to the published Walter Murray manuscript.

One account tells of a South African engineer who was crossing over the hills between the Pyke and the Dart Rivers when he noticed some pipe clay which reminded him of the clay at the South African diamond mines. He returned with two friends to find not diamonds but "rubies as big as pigeon eggs, and they collected a billy-full of them". The excitement of the discovery apparently turned the head of one of the men. The South African looked up to see one of his friends about to kill the other with an axe. He immediately shot the man who was holding the axe, and the two remaining men cleared out with the rubies.[2]

Less charitable accounts tell of a ruby mine being found "about 1870" by three men, two of who murdered their mate for his share of the gems. The two murderers then travelled not inland to Lake Wakatipu, but to the coast where they were either murdered by the settlers at Martins Bay, or aroused the settlers' suspicions by explaining first that their mate had gone back east and then changed their story to his having drowned.[3]

There are a couple of versions of what happened next. One has the two men flooding the London and Continental gem markets with thousands of pounds worth of rubies, some of which found their way to Russia.[4] The other has them putting the rubies on the market a few at a time. The last of the rubies were supposed to have been sold in Amsterdam for £90,000 in 1927, after which the Dutch wondered why there were no more New Zealand rubies competing on the market.[5]

Enter Bill O'Leary, who not only searched for the mine, but was said to have rediscovered it some time before 1910, and to have revisited it in 1930.

The rediscovery claim came from a member of a 1910 expedition which was organised to try to relocate the mine. The party was made up of four Dunedin men, John McGregor who was the manager of the Otago Foundry, his son George who worked at the Foundry, Leslie Reynolds who was a civil engineer, and a bricklayer named Cummings who had a reputation as a diviner or seer. They were accompanied by Jack Aitken and William Paulin from the head of Lake Wakatipu, who were employed as packers.[6]

The expedition did not find the lost ruby mine which produced the high-grade rubies of Murray's manuscript, nor any trace of the huge wooden platform and extensive mining operations Murray and company would have left behind. Instead they returned with what Jack Aitken and Leslie Reynolds' widow later described as "garnet, and not of gem quality". This they found between the head of Lake Wilmot and the Olivine Range to the east. According to Jack Aitken's recollections 50 years later, the expedition took place largely as a result of a newspaper article about 1908 or 1909. It was based on an interview with Bill O'Leary and described the old ruby mine and the supposed murder of one of the partners.[7]

Twenty years after the 1910 expedition, Bill O'Leary wrote the following letter to Charlie Haines of Glenorchy:

Mr C. Haines, Glenorchy.

Dear Sir,

I am wrighting these fue lines to let you now how i am getting on i found the mine but could not get into it to test it the face it is in is to steep i could not get to within fifteen feet of the mouth of it found a tree marked 1872 but could not find there camp, would of went over into Cascade ice to broken and my boots sools were worn right of and i could not stop from slipping.

arrived at Okuru May 10th will go down to the coast and out the Holyford valley will pick up a few speciments for you on my road out, Dont anser this as i will be away funds is don so i am looking for help now i will leave most of my things at lak willmot my horse is verry slow and i got a heavy load. yours truly,

W.J. O'Leary,

Okuru,

South Westland.[8]

Charlie Haines was an amateur geologist and archaeologist who farmed at Camp Hill at the head of Lake Wakatipu and had apparently grubstaked Bill O'Leary for this and other expeditions. A wallet which Bill O'Leary either stored or misplaced at Davie Gunn's Deadmans hut in the Hollyford Valley was later found by Bill McLennan, a Ministry of Works employee who was taking shelter in the abandoned hut while deerstalking. Inside the wallet was a ragged slip of paper from Charlie Haines. In very clear handwriting he had carefully described the rare and difficult-to-identify mineral osmiridium, which he presumably wanted Bill O'Leary to look for, with good reason. A line at the foot of the notes about osmiridium stated that it was quoted in the *London Mining Journal* of 21 June 1930 at £21 per ounce.[9]

The mine referred to in Bill O'Leary's letter is generally thought to be the lost ruby mine. A contemporary account suggested otherwise. After sending the letter from Okuru, Bill O'Leary headed down the coast to Big Bay, where he caught up with a party guided by Eric James. In his account of the meeting, Eric James wrote that Bill O'Leary claimed to have made an important discovery. This consisted of a man-made tunnel or drive alongside a tree with the figures 187 cut into it. There was also an indistinct fourth figure which might have been a 2 or 4. According to Eric James, the discovery was made not near Lake Wilmot, but in the upper reaches of the Arawhata River, and Bill O'Leary connected it not with a lost ruby mine but with the French-man's Gold.[10]

This confusion and tendency to blur and blend the two tales has been accentuated by subsequent reports. These have claimed that as well as the date cut in the tree, another clue Bill O'Leary found was a saw wedged in the fork of a tree and pointing in the direction of the mine. One even claimed that this was mentioned in the letter to Charlie Haines.[11] This is reminiscent of the reference in one of the Frenchman's Gold stories to a cross-cut saw worked into a beech tree as a marker. The name McLean is also common to both tales.

Other embellishments have included the claim that when Bill O'Leary returned to the mine with the equipment he needed to climb up a small bluff to reach it, he was unable to remember the exact location. The reports have also argued that Bill O'Leary's finding of the tree marked 1872 is significant because the date fitted the "about 1870" date of some versions of the lost ruby mine story.[12] The opposite appears to have been the case. The lost ruby mine date of "about 1870" was added after the contents of Bill O'Leary's letter to Charlie Haines became known.

Whatever mine Bill O'Leary did find, it is unlikely to have been the mine of Walter Murray's manuscript. Murray's narrative was so detailed in all other aspects that he probably would have mentioned carving a date in a tree. That date would presumably have been 1892 or 1894, which Bill O'Leary later misread for 1872 or 1874. Murray's mine was also in the direction Bill O'Leary was heading when he reached Okuru, not where he had been.

Whether he found either a ruby mine or Murray's lost ruby mine, Bill O'Leary did claim to have found rubies, not near Lake Wilmot or the Arawhata Valley, but between Copland Springs and Lake McKerrow.[13] Others reckoned he did too. Charlie Haines' niece, Rosie Grant, remembered him bringing to her uncle's place what appeared to be small rubies.[14] One of the captains on the Lake Wakatipu steamer *Earnslaw*, is supposed to have seen rubies Bill O'Leary brought out one trip.[15] A Christchurch man who met him at Blowfly Flat in 1927 during a trans-Haast horse trip claimed to have seen him wearing a watch chain studded with rubies which he said came from "his mine".[16] Later accounts also mentioned a watch chain decorated with small uncut rubies.[17]

As with the Frenchman's Gold, there were doubters. Many people who remembered seeing the watch chain said the story was nonsense. Photographs clearly support their derision. According to one of the doubters, Glenorchy resident Gilbert Koch, "You would think if a chap had rubies and he was talking so much about them he would be able to show them to you. But he was always just on them."[18] That he found ruby-coloured garnets is likely. That he found rubies or a lost ruby mine is not.

Nor did a subsequent attempt to find the mine meet with any success. In 1968 Davie Gunn's son Murray Gunn, Dunedin journalist turned miner Brian Jackson and Christchurch geologist turned peace campaigner Owen Wilkes searched for the mine using a guide to the location based on Jack Aitken's recollections. They found neither the mine nor any trace of it.[19]

CHAPTER EIGHT

MINERS
& MOSIES

$\mathcal{B}$ill O'Leary's 1930 mine visit was made around the time of his move from South Westland to his familiar stamping-grounds in North-West Otago. Familiar too was the pattern. Prospecting in the summer until the passes became misnomers, then wintering over, usually with families at the head of Lake Wakatipu.

The main base for his summertime prospecting activities was Martins Bay and the nearby Big Bay. It was there in 1930 that he took up temporary residence with Hugh Glass and Mac McAlpine. The two bushmen were living in one of Davie Gunn's mustering huts while on a contract to clear a 70 acre block.[1] Davie Gunn kept his huts open for anyone who was passing, including Bill O'Leary who occasionally did some work for him.[2]

Infrequent visitors at Big Bay included *Weekly News* photographer Bill Beattie, guided by Eric James on an extended horse trek from Lake Wanaka to Lake Wakatipu via the West Coast. It was after this trip that Eric James wrote of Bill O'Leary's talk of finding the mine and the marked tree.

Eric James was not the only member of the party to record the encounter. Bill Beattie later wrote:

> I was very bucked to find such a colourful character as Arawata Bill. He was most reluctant to be photographed, getting into a real tizzy, flapping his hands and shuffling his feet. "No," he said, "I don't like it. It's a lot of damn nonsense." However, he quietened down and I got my pictures. One was a close-up of him beside Dolly, his hack-cum-packhorse. Dolly had been with him for many years and they were great friends. There was

another picture of them moving over the boulder-strewn beach. When I asked to get a shot of him with his treasured iron camp oven, without which he never moved, he again began to simmer and even showed signs of "boiling", so I finished the job quickly.[3]

While waiting for the government's lighthouse service vessel, the steamer *Tutanekai*, to drop off supplies for the three locals and their visitors, Bill Beattie camped with the trio. He shared a bed with Bill O'Leary, who assured him that he neither snored nor talked in his sleep. The photographer also joined them on a 12-mile trip south to Martins Bay to see if their supplies had been landed there by mistake. Those supplies were much needed. When the visitors arrived at Big Bay, the trio were running out of food and Bill O'Leary had just returned from an unsuccessful fishing trip.[4]

To their relief, waiting for them at Martins Bay were axes, spades and tools, flour, butter and tinned food. According to the Auckland photographer, Bill O'Leary "had a glint of joyful anticipation in his eye at the thought of a royal feast" which included "stewed apples and rice, and thick ships' biscuits heavily plastered with butter and honey". The butter was the first the trio had tasted for more than two months. Well-fed and re-stocked, Bill O'Leary packed his supplies on Dolly and headed off to do some more prospecting.[5]

Bill Beattie was the first of two professional photographers to capture Bill O'Leary during the 1930s. The celebrated Christchurch photographer Thelma Kent met up with him and Dolly about 1937. They were on their way from Elfin Bay to the Olivine River on a three-month expedition.

Thelma Kent took three famous and frequently reproduced photographs of the pair. She took two of him with hat on head, standing to the left of Dolly, and a third with him to the right of Dolly, with his hat resting on her packs. The latter photograph shows he'd lost most of the red hair of his youth. What remained was snow white, as was the full beard which in more civilised climes alternated with goatee beard, and both of which had long since replaced the large red moustache and shaven face of earlier days.[6]

It was not just the odd adventurous photographer who was coming across the lone prospector at this time. During his previous incursions into the area between Martins and Jackson Bays, Bill O'Leary had the place pretty much to himself. A resurgence of interest in mining and mountaineering in the 1920s and 1930s changed all that.

There was regular contact with miners such as those who were being subsidised by the government to search for gold during the Depression. A party of four miners, including Edward Martin, who were prospecting at Martins Bay, met him and Dolly there and on the track to Elfin Bay.[7]

Colin McEwan and his mining mate Jack Swale also came across him from time to time in the Pyke, Lake Alabaster, Lake Wilmot and Red Hills area. He told them about his campsites and caves, including one near Simonin Pass which they were happy to share. Colin McEwan described it as a good place to stop, dry, with water at the back door and plenty of firewood handy. They also shared some of Davie Gunn's huts, including the Pyke hut near Lake Alabaster where they stopped for three days waiting to cross the flooded Lower Pyke River. They finally called on his West Coast ferryman's skills to take them across.[8]

Miners such as Colin McEwan and Jack Swale spent a good deal of time in the hills. It had as much to do with the overpopulation of the coast and river flats as it did with the quest for gold. These areas were crowded not with fellow miners but with sandflies and mosquitoes. They were a chronic problem. The only real solution was to clear out, which the Depression miners frequently did, into the Red Hills, to get some relief.[9]

Away from the hills there were few places you could go to escape the endless attention. The smoke of a fire or the inside of a darkened hut provided some respite from the sandflies. Outside, the so-called bushman's friends, the fantails, did their best but were hopelessly outnumbered. Clothing, including bonnets made from flour bags, provided some protection, but sandflies were known to bite through the lace holes of boots. At night their nocturnal substitutes the mosquitoes would take over the feeding duties.[10]

Bill O'Leary on the banks of the Lower Hollyford River at Martins Bay in 1931.

Some miners rubbed on pipe clay, or citronella, or a combination of oil and detergent. At night they might burn sulphur to try to keep the mosquitoes at bay. Fulltime miner and part-time wag Colin McEwan reckoned Bill O'Leary had a more effective remedy for mosquitoes. He lay on one side until they were about two inches thick all over the top of the blanket, then rolled over on the blanket, crushed them and rolled back again.[11]

There were other more welcome visits. In August 1931, two Southland Tramping Club members, Lindsay McCurdy and Jack Dobbie, also came across Bill O'Leary at Martins Bay. First they saw signs of his presence when they went to check on the rowing boat which was sometimes kept at Lake McKerrow. Instead they found two of his flax fishing lines. When they arrived at Martins Bay the following evening, they were greeted by Bill O'Leary, Davie Gunn and Malcolm McKenzie. The trampers spent two days at Martins Bay, dining on pigeons. Both admired the way Bill O'Leary, still tremendously fit in his mid-sixties, could jump onto his horse from ground level.[12]

Some people encountered Bill O'Leary once and briefly. Others came into regular contact with the roving prospector. Gordon Speden first met

Bill O'Leary catches up on the news with miners Kenneth McAlpine (left) and an unknown companion (centre) during the early 1930s.

him in 1930 near the Hidden Falls hut and later at the Pyke and Barrier huts. He also came across him in 1931 camping in a small tent alongside the temporary Public Works Department hut at Sunny Creek.[13]

It was used by men working on the ill-fated Hollyford Road, which was intended to link up with the communities of South Westland but which petered out a few miles down the Hollyford Valley. Bill O'Leary became an occasional sight for these men[14] and for men such as Vic Yarker who were working on the more successful Milford Road between Te Anau and Milford Sound. Vic Yarker first bumped into Bill O'Leary at the Marian Corner about 1935. He was riding one horse and leading another which was loaded with canvas hose pipe he'd packed in from Mossburn. He was heading down the Hollyford to do some sluicing.[15]

When Lindsay McCurdy and Jack Dobbie arrived at Martins Bay in 1931, the two trampers had not seen anyone since leaving Elfin Bay five days earlier.

The men at Martins Bay had not seen anyone for five months. They were keen to hear about the Depression, the unemployment and the dole.[16] Not that Bill O'Leary was easily disturbed by the latest happenings. News which others thought he might find a little bit shattering he would take in his stride.[17]

The isolation of the place is further illustrated by a story Bill Beattie told of arriving at Big Bay to find an unconcerned Bill O'Leary only 10 days out in his reckoning of the date. His reliable old pocket watch ensured he always knew the time of day, but he soon lost track of the day of the week and the date of the month. It was invariably one of his first questions when he met up with someone. According to Bill Beattie, the First World War had been in progress six months before the news reached Big Bay.[18]

It was easy, however, to overstate the case. One incredible story tells how:

In the mid 1930s some Dunedin blokes met up with him and they asked him if he was interested in the outside world. "Yes, I'm always interested," came the reply. "By the way, is the War over?" Well, this was about 15 years after the end of the Great War, and they said, "Of course the war is over." "Did we win?" asked Arawata Bill. "Yes we won," they told him. "Queen Victoria will be pleased," came the reply.[19]

Tales such as this contributed to the portrayal of Bill O'Leary as a hermit, a recluse who went out of his way to avoid people. It's a view for which there is some anecdotal evidence. Arthur Hamilton told of how he and his brother Wilf had been at Lake Howden hut with parties of tourists and had known Bill O'Leary to pass by in the night to avoid the occupants. In Arthur Hamilton's view, "Any normal man would have been into the hut like a shot to meet people after so long in the bush, but not Bill."[20]

Similarly, Davie Gunn told of the time some young trampers arrived at Martins Bay. When Bill O'Leary heard they were coming he took his gear and Dolly and rode across the Hollyford River to the old McKenzie homestead

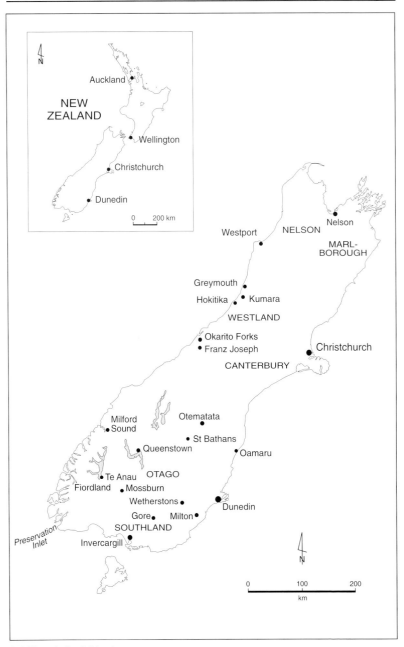

Bill O'Leary's South Island.

Arthur and Wilf Hamilton bringing a party of tourists ashore at the mouth of the Greenstone River. Bill O'Leary is lending a steadying hand at the far side of the boat.

in the distance. When one of the trampers asked if Arawata Bill had gone, Davie Gunn told them he had but he'd be back once they had departed.[21] Fergie Heffernan reckoned Bill O'Leary would have nothing to do with the tourists, wouldn't go near them, wouldn't speak to them, wouldn't answer them.[22]

It's not a prevailing view. Most said he was quietly spoken, retiring, shy with strangers, but nevertheless pleased to see people. He was at home in their company, very friendly, made them welcome when they arrived at a hut. He spent the greater part of each year by himself not because he was unsociable, but just because he preferred it that way. He was a loner but never lonely. He enjoyed his own company and solitude away in the hills for months at a time. He felt little need for the regular company of others. He was staunchly independent, wanted to be his own boss, liked the sense of absolute freedom.[23]

Even Davie Gunn was quick to add that Bill O'Leary was no hermit. He would go out of his way to avoid people he didn't like, but he was easy to know and was pleased to see people he did like.[24] Bill O'Leary's willingness to lend a hand and be photographed with tourist parties supports the view.

CHAPTER NINE

GUMBOOTS
& GLACIERS

*J*ust as the increase in mining activity brought more people into Bill O'Leary's territory, so did the revival of interest in climbing. It was these direct and indirect contacts which more than any others were responsible for the creation of the legend of Arawata Bill.

One of the most important contacts occurred when two generations of explorers came together at those relics of an earlier and more ambitious search for gold, the Dredge huts in the Dart Valley. Two members of a New Zealand Alpine Club party, Jack Holloway and Chid Sealy, met Bill O'Leary at the huts in January 1937 during the party's three-month exploration trip.[1]

The prospector and the two Otago University students were held up by several days of rain and snow in what was an extremely bad season. Chid Sealy later wrote:

> ...three days were very pleasantly spent in the company of the well-known veteran, Arawata Bill, who was also weatherbound at the huts on his return from an unsuccessful attempt to cross into the Joe River by means of O'Leary Pass. Many an interesting tale of early experience was told by this hardy old-timer of nearly 80 years of age as the three men sat around the fire and watched the billy boil, while outside the rain poured incessantly, and the Dart, which had to be forded before further progress could be made, rose steadily higher.[2]

Jack Holloway's account of the meeting gives a more revealing insight into Bill O'Leary's activities, and into the effect heavy snowfalls could have on a trip which Bill O'Leary in better weather had previously done in a day:

> This hardy old prospector, between 70 and 80 years of age, had been
> turned back on an attempt to cross over into the Joe River. For several
> weeks he had been engaged in laboriously digging steps with a pick-axe
> for two thousand feet up a steep snow-covered mountain side to the
> summit of O'Leary Pass. Up these snowfields he had relayed his food and
> mining equipment, but had then found himself unable to descend the
> other side under the snow conditions then prevailing. Accordingly he
> carried everything back down the Dart once more, and was about to set
> out on a long ride to the West Coast, where he might reach his objective
> from the other side.[3]

Bill O'Leary was 71.

*Chid Sealy (left) and 'Arawata Bill' at the Dredge
huts in January 1937. The bad weather let up long
enough for Jack Holloway to take the photograph
of his climbing partner and the "hardy old-timer".*

In a subsequent account, Jack Holloway said Bill O'Leary's only concession to mountaineering technology was not a pick-axe but a long-handled shovel with which he would dig deep steps up a snowfield. In this later account, Jack Holloway also referred to the prospector's wonder at first seeing a pair of crampons. He immediately wanted to know what they were and where he could get a pair, because he always wore thigh gumboots, even when in camp, and "slipped horribly on the snow".[4]

Similar hearsay stories were also related of a party of six students meeting Bill O'Leary either on Cosmos Peak to the west of the Dredge huts, or climbing a glacier in the Barrier Range. The climbers with their crampons and ice-axes were amazed to find him in his thigh gumboots with the tops rolled down, and cutting steps in the ice or snow with either a short-handled or long-handled shovel.[5] One account had him completing the impossible by crossing the pass "through snow and ice, in his thigh gumboots with his old horse for company".[6]

It's unclear how many times Bill O'Leary crossed over O'Leary Pass. Jack Holloway said it was only once as far as he could tell, from the Joe to the Dart, although he could never find out by what exact route.[7] This ignores the first return crossing he made from the Dart to the Joe and back in 1897, and probably others as well. Fergie Heffernan reckoned Bill O'Leary crossed the pass "quite frequently".[8]

In his subsequent account, Jack Holloway, as guest speaker at the Massey University Alpine Club's annual dinner in 1976, enthused a new generation of climbers about the man he had come to admire so much:

> He was a kind, gentle, shy man who would do anything for anybody at any time, no matter how long it took. On one occasion he was worried that we did not have sufficient food for a projected trip from the Dart River across the Dart, Barrier and Olivine Ranges to Big Bay and, sure enough, we were out of food by the time we reached the coast. But Arawata was already there, with more food, after a three week trip down the Dart, up the Greenstone, down the Hollyford and up the Pyke on the off chance that he would meet up with us. He would accept no thanks because it was no more than anyone would do for anybody else in like circumstances and did not require mention.[9]

Jack Holloway had known Bill O'Leary since the climber was about 12 years old on a family summer camping holiday at Elfin Bay in the mid-1920s. For several years he would meet the prospector occasionally for a cup of tea and a chat at his camp a few miles downstream from the old Rees River

Albert Jackson took this photograph during a 1935 expedition, looking down the Dart Valley to O'Leary Pass (centre right) in the Barrier Range.

bridge at the head of Lake Wakatipu. Later he would meet him regularly at such places as the Dredge, Lake Howden and Deadmans huts.[10]

As well as whetting the young man's appetite for climbing, Jack Holloway later acknowledged that from Bill O'Leary he had heard much of the Olivine country behind the Barrier Range, "though his accounts were at times confusing and difficult to follow".[11] It was not what Jack Holloway, the keen map-maker, regarded as real information:

> We would tell him where we had been, and sketch maps, and only then would he fill in with bits of detail and recollections of his visits there. He had his own names for some places, mainly rivers, and/or lakes, but usually no names at all, and he certainly could not draw even a rudimentary map.[12]

Bill O'Leary made no maps it's true, but climbers who followed received

ѕenefit from the route markers he had made for himself, in the
. blazed tracks in the bush and stone cairns above the bushline.

ₙ the mid-1930s a party which included Jack Holloway and Albert
Jackson followed a blazed trail on the Dart side of O'Leary Pass, which had
possibly been cut through the bush by Bill O'Leary within the previous two
years.[13] About the same time, a Southland party stayed at a mai mai Bill
O'Leary had built about a mile up the Barrier River from the Pyke Junction,
and then followed a trail he had blazed up through the Barrier Gorge.

A member of the party, Alex Dickie, later wrote:

> The trail rose steeply up along the hillside, and the river below on our left
> seemed to come down in one continuous cataract from regions beyond.
> It was now that we began to feel somewhat ashamed of ourselves as we
> laboured along under those packs, for we thought of the hardihood and
> endurance of Arawata Bill, who in years is approaching the three score
> and ten. He first cut this track and then made numerous trips along it
> with loads on his back on his way to his happy hunting grounds in the
> Red Hills, and it was the thought of this that made us feel small indeed.[14]

The same trail had been followed by Jack Swale a couple of years earlier.
According to the miner, the sides of the gorge were so steep in places that it
was impossible to get round without some assistance. Bill O'Leary's solution
was to cut poles 10 to 14 feet long and lay them between the growing saplings
to make a platform. This he had done for several chains.[15] The trail was later
located and re-blazed by Jack Holloway in the 1930s[16] and by a party from
the Otago Tramping Club in the 1940s.[17]

In 1953, a party from the Hutt Valley Tramping Club, including Arnold
Heine, followed Bill O'Leary's footsteps from the Dart to the Joe over O'Leary
Pass. They were guided by several of his old markers down the Joe and avoided
having to ford the dangerous river.[18] Two years later, the party which rediscovered his cave at the foot of the Ten Hour Gorge followed what they

thought might have been his cairns from the Olivine Ice Plateau to Williamson Flat.[19]

Bill O'Leary also left a trail of cairns over Simonin Pass between the headwaters of the Cascade and Pyke Rivers in the Red Hills Range. These marked the best route "past the bad places, of which there were many" for John Pascoe and fellow members of the Canterbury Mountaineering Club in 1952–53. They also came across a rusty shovel and a hut "rotted into a heap of fragments". [20] This casual encounter would have crucial importance in the making of the Arawata Bill legend.

In addition to these marked trails, Bill O'Leary had his own route from Lake Wakatipu to the Coast, which took him from the Dart Valley up the Rock Burn and over a high pass into Hidden Falls Creek and the Olivine River.[21]

Climbers also came across signs of his prospecting work, such as a mound on the Red Hills which he had dug out in his hunt for gold. They made use too of his old campsites, such as one at the head of Lake Alabaster where he would pitch his tent and use as a summer base.[22]

In the absence of evidence, climbers even deferred to the possibility that he might have been there before them. Arnold Heine, for example, wrote that while his party's traverse of Cilicia Peak, of one of the Five Fingers and of Camp Oven Dome high above Williamson Flat appeared to be the first ascents, it was possible that Charlie Douglas or Bill O'Leary might have made unrecorded ascents of the peaks.[23]

Bill O'Leary was already well-known in South Westland and North-West Otago. His disappearance in 1929 had received some wider publicity. As more and more climbers came across him during the 1930s, his fame began to spread.

While Bill Beattie did not publish his account for another 40 years, Eric James wrote contemporary reports of their trip, which were published in the *Otago Daily Times*. They referred to Arawata Bill, the well-known prospector, as though it was his real name.[24] Contemporary articles by Jack Holloway and Chid Sealy also appeared in the daily press in Dunedin and Christchurch, including Jack Holloway's remarkable gumboots and pick-axe story. Both Chid Sealy and Jack Holloway used his nickname alone.[25]

They and climbers such as Alex Dickie and Albert Jackson also mentioned him in written accounts of their trips which were published in the widely-read *New Zealand Alpine Journal*. Again he was mainly referred to as Arawata Bill, although sometimes his real name was added. It was the same story during the next three decades of write-ups in the *Alpine Journal*.[26] Similarly, when Lindsay McCurdy completed his trip book at the end of his expedition, he only referred to Arawata Bill. More than 40 years later when he provided an account of the trip for publication he still only used the nickname.[27]

Publicity also came in the daily press in the 1930s as a result of Bill O'Leary's odd trips to Dunedin, where it took him a day or two to adjust to the street traffic. After one visit, an *Evening Star* newspaper columnist observed, "It is seldom that Dunedin has the opportunity of entertaining a real live hermit of such world-wide distinction, for Bill is just as well-known overseas as locally."[28]

The reference to Bill O'Leary's worldwide distinction was no mere hyperbole. In 1935 an article by its Dunedin correspondent, Harry Fortune, appeared in the London-based *Wide World Magazine*. In its day the magazine produced ripping good yarns from around the globe. The article was based on second-hand information from the writer's mountaineering friends who had encountered Bill O'Leary in the Hollyford Valley. Referring to him only by his nickname, the article painted a romantic picture of a character called Arawata Bill fossicking for gold among the isolated and beautiful valleys for up to eight months at a time, making his home wherever he chose under the wide and starry sky, monarch of all he surveyed. For good measure it threw in a version of the legend of the buried sea-boot full of gold and his obsession with the desire to find it.[29]

The odd signs of his presence were enough to indicate that Bill O'Leary thoroughly covered most of the country between the Lower Hollyford and the Haast rivers. Tales were told of expeditions further afield, such as a voyage supposedly done with Davie Gunn by open boat from Martins Bay down the coast to Preservation Inlet.[30] It's a story which was doubted by Davie Gunn's son Murray. His father would never have taken the time off work and neither of the men could swim a stroke and were unlikely to risk their lives in an open boat off the rugged Fiordland coast.[31]

Jules Berg,[32] the Swedish seaman-turned-prospector who became known as the hermit of Preservation Inlet, claimed that he twice travelled overland from Preservation Inlet to Milford Sound with Bill O'Leary. In the same breath he claimed that during one of those trips he saw three supposedly extinct moa, and that they both saw the huge footprints of the birds on the banks of a stream.[33] It's the only suggestion that Bill O'Leary ever travelled overland south of Milford Sound and is also doubtful.

Perhaps in jest, Davie Gunn told of a shorter trip Bill O'Leary made. Davie Gunn had cut a deviation from the original track at the head of Lake Alabaster. It was dark when Bill O'Leary reached it. He started on the new track, got back on to the old one and walked round and round in a circle without realising it. There happened to be a dead bullock lying on the track with a few dogs in attendance, which he passed several times. When one of Davie Gunn's workmen finally gave Bill O'Leary his bearings, his first words were supposed to have been, "There are a lot of dead cattle on the track, and there are dogs at every one."[34]

For all his exploits, Bill O'Leary was a cautious traveller. He had to be to survive in such rugged country for so long. If he was by himself and a river was too high or the weather was too bad, the non-swimmer would sit tight in his tent or under a rock shelter until it was safe to cross or the weather cleared. If he had Dolly, the first class swimmer, he would be more confident. Instead of looking for a ford he would plunge into the river, horse and all, and the current would sweep them to the other side.[35]

Caution was no complete protection from trouble. On several occasions he had alarming experiences. He later spoke of an incident in which he tumbled over a 70-foot cliff. The fall should have killed him. It didn't but he was severely injured. There was no one to come to his aid. He had a hard struggle to climb back up the cliff, mount his horse and painfully ride 35 miles or more to the head of Lake Wakatipu. From there he was taken by launch to Queenstown and on to Dunedin Hospital, about 230 miles away.[36]

On another occasion, in the winter of 1936, he was taken from a mining camp near Skippers, up the Shotover River, to a private hospital in Dunedin for a "severe operation". According to one story, when he came to, he got out of bed and went looking for a toilet. When he was near the toilet door he heard a nurse screeching down the corridor, slammed the door and locked himself in. When he was ready to come out, he found the matron and several nurses waiting to carry him back to bed, but the "tough old bird" walked. The matron reckoned "he should be dead if he had been an ordinary man." Another version has him being returned to his bed after a hospital attendant had found the pyjama-clad patient wandering the corridor looking for his clothes "with the object of making a break for open country."[37]

A further incident he recalled took place during a big flood when two lakes about eight miles apart, probably Alabaster and Wilmot, were practically joined by floodwaters. He managed to send Dolly to safety but the small ridge where he was camped became surrounded by raging water right up to the tent.

For three days he was trapped on a narrow strip of land without food. He kept himself alive on eels, which he caught with great difficulty with a gaff. He was more than 50 miles from any settlement and if he had perished in the flood, no one would have known.[38]

As with other stories about him, this one grew in the retelling. One version has him trapped not for three days but for three weeks, surrounded by floodwaters and living on eels and three loaves of camp oven bread. When the floodwaters receded he rode out to the Hidden Falls hut, where he arrived looking like a skeleton.[39]

A less dangerous but momentarily just as alarming incident was related by Alice McKenzie, based on a story she had heard about him:

He had been away for some time on one of his trips and was so long overdue that it seemed he must have passed away. Mr Cashmore was going into that country and the police told him that if he came across the body of Arawata Bill he should cover it up, but not bury it, so that the police could make an inspection. Away up in the Upper Pike valley, somewhere near Lake Wilmot, Mr Cashmore came across the old man's camp. There stood his tent, with half the length of Arawata Bill's legs protruding. Mr Cashmore thought he had better pull the body into the tent and leave it there, so he opened the tent and took hold of the corpse. To his astonishment the "corpse" sat up with a start, and it would be hard to say which of the two was the more startled. It was broad daylight and Mr Cashmore had been expecting to find a dead man, so his mistake can be easily understood.[40]

There is even some doubt over this often-repeated story. A less dramatic account of probably the same incident was told by Jack Swale. Bill O'Leary had been delayed by weather up the Pyke River and two Davie Gunn employees, Tip Cashmore and Chris Hooker, thought he must have taken ill and went to see what the story was. They headed up the Pyke and met him coming down off the Red Hills. He was running short of food and was pleased to see them. As a way of repaying them for their efforts, he told them about a prospect he had, but they couldn't find it.[41]

Bill O'Leary wasn't always the cautious traveller. He once took a tumble conducting an experiment coming down the snowfield from O'Leary Pass to Cattle Flat in the Dart Valley. He thought it might be a time-saver and a bit of fun to sit on his tin gold panning dish and have a slide. He lost control and finished up end over end on the rocks at the bottom. He is said to have lost "yards of skin" and to have been "a pretty scaly looking object when he arrived".[42]

That attribute was to become part of his legend. In the words of Jack Cron junior, "he roughed it and he always came out alive. That was the big thing."[43] Similarly, from Frank Heveldt, "He was a man who always arrived. He knew where he was heading and he always got there",[44] even if at times a little the worse for wear.

CHAPTER TEN

WAKATIPU
WINTERS

*W*hereas Bill O'Leary spent the summers in the hills, emerging only when his provisions ran out, winter and old age saw him looking for shelter and work with miners at a camp near Skippers,[1] or with families at the head of Lake Wakatipu. Perhaps out of a desire not to overstay his welcome, he would spend a winter or two with one family, then move on.

Around 1930 he and Dolly turned up at Cooks' Sawmill at the mouth of the Route Burn, about 10 miles from the head of the lake. He looked thin and hungry after his summer prospecting based at Martins Bay. There was no work for him at the mill but owner Neil Cook gave him a job as cowboy, gardener and rouseabout. Cowboy was the romantic title for the mundane task of milking the house cows. He also looked after a half-acre of garden which supplied the Cook family and the mill workers, and he fed the pigs and kept up a supply of kindling and firewood.[2]

During his two to three-month stay he lived in the family homestead with Neil and Lenore Cook and their six children, Neil, Charlie, Wattie, Bernie, Lenore and Janet. Dolly spent the winter with the 20 to 30 mill horses and dined on chaff and the daily hen's egg the teamster would break into it to make her coat shine.[3]

With the coming of the warmer spring weather and with a few bob in his pocket, Bill O'Leary would get itchy feet. He would pack his food and belongings on Dolly and head back into the hills for another extended period of prospecting. When the snow started falling he would re-appear, to be greeted by the Cook children who would run up and ask if he had brought out the rubies or diamonds he had talked about the previous winter.[4]

As always, he got on well with the children, whom he delighted with stories about his adventures. Neil Cook junior recalled:

> He used to sit on the side of the bed at night and tell us stories about what happened in there. Crossing the lake how the boat would go down and he would have to swim to the shore and he would lose everything and the next day he would go back and fish the boat out. He used to tell us all these stories which we thought were wonderful but I think he was pulling our legs a lot of the time.[5]

After a couple of seasons with the Cooks, Bill O'Leary began spending the winters at Elfin Bay Station with Gi and Annie Shaw and their daughters Betty and Elfin. He earned his keep partly by planting out and tending the vegetable garden. As he was not there when the vegetables were ready, Annie Shaw would preserve peas for him to eat when he returned. Sometimes he would arrive with three or four pigeons cooked in his portable oven saddle bags. Annie Shaw would supplement the pigeons with peas and potatoes for him and the Shaw family to tuck into. When he left it would be with one of her fruitcakes packed away with his supplies.[6]

M.A. GUNN/CRAIG PRINTING

Bill O'Leary at Elfin Bay on the western shores of Lake Wakatipu.

As well as looking after the garden at Elfin Bay, Bill O'Leary milked the cows, did odd jobs and helped keep up a great pile of firewood and a full woodbox. It was a complicated business. Manuka and beech trees were felled with a two-way saw, the logs dragged by horse and chain to the kerosene-motor, belt-driven

Bill O'Leary posing outside his hut at Elfin Bay in 1932.

sawmill and the sawn blocks split with an axe. The shorter lengths were destined for the kitchen stove, the longer ones for the open fires. It was the shorter lengths which made Bill O'Leary a well-intentioned liability in the Shaw kitchen. He was always adding wood to the stove fire. He couldn't help it. It was a habit which came from decades of frequently feeding hungry camp fires. The result was overheated hot plates and oven, and burnt food.[7]

Bill O'Leary had his own hut at Elfin Bay, but he would spend the evenings with the Shaw family in front of the big open fireplace in the sitting-room. He would tell the children stories and sing and dance around like a child whenever music was being played. After the children had gone to bed the adults would play cribbage late into the night by the light of kerosene lamps and candles.[8]

The Shaw children thought Bill O'Leary was Father Christmas, with his white beard and the gifts of lollies which Elfin Shaw remembered him always carrying in his old canvas raincoat:

> He had two brown paper bags full of blackballs and peppermints. The blackballs were turning white with age and the peppermints were black with the many months and many miles he had carried them. But we didn't care. He was very fond of us kids and we thought he was marvellous.[9]

Father Christmas was one of the more flattering descriptions of Bill O'Leary's appearance. It ranked with portraits of him looking like one of the

The Shaw girls riding their pet heifers at Elfin Bay in 1934. From left: Elfin Shaw, Bill O'Leary and Betty Shaw.

Apostles with his white beard, beautiful complexion and clear or sparkling or bright twinkling blue eyes.[10] Others were less flattering. Their blunt descriptions included sandy complexion, ferret-like eyes and toothless. His hooked nose made him look more like a morepork or ruru, the native owl. He had a little parrot face with a big nose which looked like a parrot's beak buried in a long bushy white beard.[11]

Of the two Shaw children, Elfin was his favourite. She would traipse around after him, catch and feed Dolly for him, go for rides and watch as they performed tricks. In one, Dolly would turn her head and give him a nudge in the backside to help him into the saddle. This was partly to amuse the children and partly because age was making it more difficult for him to complete the exercise unaided.[12] For purely practical reasons he had taught her to wriggle under fallen trees too high to step or jump over while they were going along bush tracks.[13]

In many respects Dolly was an equine version of her master. She was quiet and gentle. She preferred to be away from the crowd. In a paddock with other horses she wouldn't mix. She was content to head off into the hills whenever he was ready. It was suggested that she even took a keen interest in his prospecting activities. At a campsite she could be let go and relied on not to stray. If he was going off somewhere he couldn't take a horse, Dolly would pack in gear as far as she could, and then be left to graze until he returned.[14]

Local legend has it that things didn't always go according to plan. Bill O'Leary is said to have told of the time he and Dolly stopped the night at the head of Lake McKerrow. When he went to get her in the morning, she wasn't there. She had cleared out and made her way down to Martins Bay. This involved a swim of at least four to five miles, to get round a blocked-off part of the lakeside track, and maybe the entire length of the lake. The attraction at Martins Bay was said to be one of Davie Gunn's stallions.[15]

Bill O'Leary's fondness for Dolly was in sharp contrast with his attitude towards other animals around the station house at Elfin Bay. For some reason he had developed a dislike for dogs. He would hiss at the chained farm dogs and feign to hit them with a stick.[16] This dislike apparently stopped short of cruelty. One time he found one of Davie Gunn's dogs struggling alongside Lake Alabaster. It had strayed and jammed its head inside a syrup tin. He felt sorry for the dog and took it all the way back to Davie Gunn's camp.[17]

At Elfin Bay he would also provoke the pigs in the pigsty and was not fussy about where he tipped their bucket of skimmed milk. This backfired on one occasion when he teased a big tusked boar called Billy. It went berserk, broke out of its sty and chased him round and round his hut. When he took refuge inside, Billy waited for him on the doorstep. The boar had to be shot.[18]

During one of his return trips to winter over at Elfin Bay, Bill O'Leary encountered a South African tourist who would give him one of his first tastes of international publicity. Pamela Bourne, the daughter of Sir Roland Bourne, the Secretary of Defence in South Africa, was on a world tour with her mother, and the pair were staying with the Shaws. She first met Bill O'Leary at the Rat's Nest hut up the Greenstone Valley, where the South Africans were roughing it for a few days, and he was passing by on his way to

Elfin Bay Station. They then stayed together with the Shaws. Pamela Bourne included a few pages on 'Arawata Bill' in her account of the two-year adventure, *Out Of The World*, which was published in London in 1935. Referring to his quest for the legendary sea-boot full of gold, she described him as "that rare phenomenon, a practical romantic", and wrote that "he might have been Don Quixote returning from his wanderings." She also added another chapter to the continuing story of Bill O'Leary's returns from the dead, ending her account by reporting that she had heard he had died that winter.[19]

After a few seasons at Elfin Bay, a very much alive Bill O'Leary wintered over with Tom and Ivy Bryant and their sons Harry, Tom and Ernie at the former Glacier Hotel at Kinlock. He arrived there one May and asked if he could pitch a tent on their property. They gave him his own hut, Old Bill's hut, as it became known. He slept there and cooked on an open fire. He had his main meal with the family and spent the evenings with them, chatting about his wanderings and indulging his passion for music. Tom Bryant senior playing the *Highland Schottische* on his accordion was guaranteed to have the old prospector jumping up with a whoop and kicking up his heels.[20] When there was no music to be heard, he made his own. It was not unknown for

M.E. WEBB

Locals and tourists waiting for the steamer at Elfin Bay wharf in 1934. From left: two unknown tourists, Annie Shaw, unknown tourist, Fred Dollimore, Moira Crowe, Bill O'Leary, Elfin Shaw, Betty Shaw, Gi Shaw, unknown tourist.

Bill O'Leary and Dolly returning to Elfin Bay Station via the Greenstone Valley after a summer prospecting trip.

him to be plodding along on a laden Dolly, while waving a stick to some silent tune in his head.[21]

In return for the Bryants' hospitality, Bill O'Leary lent a hand with the usual jobs. He took care of the firewood, dug and planted out the garden and helped with whatever else needed doing about the place.[22] He also performed the same tasks for Huntly and Dorothy Groves, who were raising their three children Norrie, Shirley and Joan at Woodbine Station above Kinlock. His reward was a bed and tucker and money for supplies when he was finished.[23]

Further on up the road at Routeburn Station, lived Huntly Groves' brother Lewis and his wife Nellie and daughters Reta and Betty. Bill O'Leary would visit the family on his way to Kinlock to collect supplies brought up the lake on the *Earnslaw*, and Nellie Groves would give him a meal.[24]

He would also have the odd meal at Paradise House, the guesthouse run by the Aitken family, including Jack Aitken's nephew Fergie Heffernan and his wife Barbara. Not one for sitting about the table at the end of the meal, Bill O'Leary would soon be away outside chopping wood. He liked to be doing something. Unless Barbara Heffernan happened to be playing the piano, often accompanied by Fergie Heffernan with bow in hand, playing the saw. As long as the room was filled with tunes such as *I'll Take You Home Again Kathleen* or *The Mountains O' Mourne* or *When Irish Eyes Are Smiling*, Bill O'Leary was happy to sit and listen.[25]

There were other calls to be made too. A visit to Charlie Haines at Camp

The steamer Earnslaw *plying the still waters of Lake Wakatipu in 1939.*

Hill where he might stay a few days or maybe a week or two. A call on Charlie Haines' brother-in-law Bill Grant, who had a sawmill at Kinloch.[26] A trip to the post office at Glenorchy to collect his old age pension.[27] A wander up the road to say gidday to miners such as the Sharpes with whom he spent one Christmas in an annex at the side of their house at Cosy Dell.[28]

At Elfin Bay, Kinloch and Glenorchy he would also join the locals in greeting one of the government-owned New Zealand Railways steamers which serviced the lakeside settlements. Early on there was the wooden paddle-steamer *Antrim*, until it was scrapped in 1920, and the paddle steamer *Mountaineer*, until it was withdrawn from service in 1932. That left the screw steamer *Jane Williams*, later renamed *Ben Lomond*, which would continue to operate until it was scuttled in 1952, and the twin-screw steamer *Earnslaw*. "The Lady of the Lake", as she was known, arrived from Queenstown Mondays and Fridays during the winter and Wednesdays as well in the summer.[29] Once the steamer had been secured to the wharf, one of the deck-hands might say to the disembarking tourists "That's Arawata Bill over there" and they would go over and talk to the celebrity and snap his photo.[30]

During his wintering-over at the head of the lake, Bill O'Leary began a

C. DE VRIES

Bob Moodie (left) and Bill O'Leary enjoying a fine day on the waterfront in Queenstown. Tied up in the background is the Earnslaw.

correspondence with Bob Moodie, whom he had met and taken a shine to while Bob Moodie was working on farms and tracks around Lake Wakatipu and the Haast Pass, and hunting, tramping and fossicking in the region. Along with his note to Charlie Haines, the newsy letters to Bob Moodie are the only known examples of Bill O'Leary's correspondence to have survived, and endorse earlier comments about his lack of formal education.[31]

C. DE VRIES

One of Bill O'Leary's letters to his mate Bob Moodie. Transcripts of this and other correspondence are reproduced in the Appendix.

Among the families with whom he stayed at the head of Lake Wakatipu, Bill O'Leary was remembered as being neat and tidy and clean and respectable. This distinguished him from the majority of swaggers who appeared in the district, particularly during the Depression. He was forever washing himself. He would come in for his evening meal scrubbed and dressed in decent clothes. His snow white beard and what was left of his snow white hair was neatly trimmed. At Elfin Bay, Gi Shaw doubled as the resident barber.[32]

Bill O'Leary's clothes received as much attention as he did, although no amount of washing could get rid of the not unpleasant waft of campfire wood smoke. Any mending or patchwork was neatly sewn and nice and square.[33] It had been the same over the West Coast where, in the words of Jack Cron junior, "he never went round in Old Pat's clothes" and always had a good suit to put on for an occasion.[34]

Occasions included arriving at a settlement. Fergie Heffernan once came across shreds of Bill O'Leary's old clothes hanging out to dry on a tree branch up the Dart Valley. He had taken a bath in the creek and changed into his best clothes so that he would arrive "spruced up to the nines".[35]

The personal habit of cleanliness was reinforced when he began to suffer from a skin complaint known as beech-itch or birch-itch. The common belief was that it was caused by pollen from beech trees, particularly when the bush was wet and anyone who was liable to be affected brushed their skin against the damp branches. Bill O'Leary certainly believed that. He claimed the only way he could avoid the itch was to avoid the beech trees. It was the reason he chose a patch of manuka scrub as his campsite near Lake Rere, close to the junction of the Greenstone and Caples Rivers.[36]

The heat of a summer's day and a winter's fire would aggravate the condition. In the back country it would sometimes force him to take to his tent. This has been offered as an explanation for his legs sticking out of his tent in the cooler air during the reputed Cashmore incident.[37] In a hut he would scratch all night. At the Shaw house at Elfin Bay, soaking in a hot bath of Epsom salts provided temporary relief.[38] He also resorted to folk remedies such as drinking so-called beech beer or birch beer made from rain water trapped in rotten logs, and rubbing on the sticky orange jelly derived from split flax leaves.[39]

Despite the relief Bill O'Leary claimed from avoiding beech trees, the condition may well have been caused by a dietary deficiency, possibly through a lack of fruit and vegetables over an extended period.

CHAPTER ELEVEN

FAMILY
& FLYING

$\mathscr{B}$y the late 1930s Bill O'Leary's summer expeditions into the back country were all but over. In 1938 he moved down Lake Wakatipu to live with the Thompson family near Queenstown. Bill and Elsie Thompson, later mayor and mayoress of the tourist town, had 300 acres at Fernhill, including a strawberry and pip fruit farm. They fixed him up with a hut near the family homestead, where he stayed for about a year. He then moved into a more isolated and private hut at the Three Mile Creek, where he spent another year. Both huts had built-in beds. Bill O'Leary slept on the floor, wrapped in a grey woollen blanket.[1]

The 72-year-old came to the same comfortable arrangement with the Thompsons as he had with the families at the head of the lake. In return for

Ploughing at Thompsons' strawberry gardens at Fernhill in the late 1930s. From left: Harry Young, Bill O'Leary and Bill Thompson with Ned.

Bill O'Leary at Thompsons' farm at Fernhill in 1938.

meals and accommodation, he worked as a rouseabout. He helped with the horses and the ploughing, picked strawberries, pruned fruit trees and milked the cows. Having his own hut gave him a sense of independence, but he ate his meals with the family and joined them in the evening for a sing-song around the piano. His favourite tune was the ever-popular *Boomps-A-Daisy.*[2]

Walking had been such a major part of Bill O'Leary's life that he could not stop. Every Sunday he would take a five-mile stroll in his tweed suit and hat to Arthurs Point beside the Shotover River, then walk five miles back. Decades of practice had perfected the art of quickly covering the ground without seeming to be in a hurry.[3]

Sunday was also church day. He became a regular attender at St Joseph's church in Queenstown, his advancing years coinciding with one of the few periods in his adult life when a Catholic church was at hand. He often went with his mate Alex Munro, who worked as a deck-hand on the *Earnslaw* for 26 years. On one occasion, Bill O'Leary took a back seat until he saw Alex Munro sitting up front. Instead of joining his friend by going down the aisle, he took the direct route and clambered over the intervening pews.[4]

Bill O'Leary's other great mate in Queenstown was Jock Edgar, who had been a prominent figure in the local tourist trade. He had bought the graceful old launch *Thelma* and built the modern passenger launch *Kelvin*, which he operated on the lake. Ahead of his time, he also developed walking trips from the head of the lake. Bill O'Leary and Jock Edgar had formed a friendship despite their differences. Unlike his prospecting pal, Jock Edgar was a heavy gambler with a habit of going on booze benders for several days at a time.[5]

There was one area of their lives Bill O'Leary and Jock Edgar did have in common. They were both confirmed bachelors. In Bill O'Leary's case, it was said that women were something completely foreign to him. They never led him astray. The crude explanation was that he didn't like women, even hated them.[6]

There's no evidence for that and plenty to indicate that he had a profound respect for women. His bachelor status was more likely to have been a symptom of his way of life and natural reticence. He seldom settled in one place for very long. Most of his life was spent alone or in the company of small groups of men. In the company of women he was initially shy and uncomfortable. Once familiarity had overcome shyness they got on very well.[7]

Bill O'Leary had spent a lifetime on foot and horseback. In pre-World War II Queenstown he took to the air for the first time. He liked to watch the aeroplanes taking off and landing at the Frankton Aerodrome, where he was befriended by Harry, later Sir Henry Wigley, the managing director of the Mount Cook Company. If Harry Wigley had a spare seat in his plane he would invite Bill O'Leary along for the ride. The old man protested that he didn't have any money and couldn't afford to pay, but Harry Wigley easily persuaded him that it didn't matter:

> I used to take him over the district he was too old to go into any longer. … We could cover in a few minutes country it had taken him days to walk through, and his eyes used to sparkle. I liked to take him with me just to see the look of happiness on his face.[8]

The image of the flying fossicker doesn't square with stories about Bill O'Leary not having a head for heights. He was said to have detested any high climbs. He would stick to the river valleys and never go to the top of a peak to have a look around or get his bearings. If he came to a fence he would rather walk around it than climb over it.[9] Once at Kinlock, he declined to ride on top of a dray load of hay. It was too high and he walked the two miles home.[10]

The flying image also clashes with stories of Bill O'Leary's earlier aversion to what he called "new fangled machines". He was said to have become very alarmed after taking up an offer of a ride in a Model T Ford owned by Claude Capell, who ran the Lakehouse fishing lodge at Lake Hawea. On the way down the hill from the Lakehouse to Hawea Flat to catch the bus to Dunedin, Bill O'Leary "held on to Claude for dear life, pleading with him to 'Stop, stop!'" Back from Dunedin, he declined a return ride from Hawea Flat to the Lakehouse. At the sound of a car he took to the paddocks and hid until the road was clear again.[11]

From time to time Bill O'Leary would leave the Thompson farm and head up the lake on the *Earnslaw*, then walk through the Greenstone to the Hollyford Valley to visit his old haunts. Murray Gunn was helping his father during the school holidays in 1938 when he met Bill O'Leary at the Hidden Falls hut. He was returning to Queenstown carrying a heavy pack in what Murray Gunn believed was Bill O'Leary's last trip through the Hollyford. Even in his seventies, he was so fit that he would run rather than walk.

The Hidden Falls hut about the time of Bill O'Leary's last visit to the Hollyford Valley in 1938.

Davie Gunn asked him to shut a gate. He ran to do the job and ran back.[12]

Two years earlier Davie Gunn had confirmed his place in the country's legend book. He had made a 22-hour trip on foot, rowboat and horseback from Big Bay to the nearest telephone 45 miles away at the Marian Corner to get help for the four survivors of a plane crash. A fifth person was killed. Tragedy later befell the rescuer. Davie Gunn and a young boy were drowned while crossing the Hollyford River on horseback on Christmas Eve 1955. They never found Davie Gunn's body.

There are various accounts of what was supposed to have been Bill O'Leary's last trip into the back country. One version has him walking into the Barrier hut and looking "really sad and down to it". He had developed beech itch and would no longer be able to go into the bush, including the Skippers Range between Lake McKerrow and the Pyke River, which he still hoped to cover.[13]

Other versions are closely tied in with Dolly. One has her failing because of old age and no longer able to undertake the trips demanded of her. On her last expedition she is said to have twice fallen over precipices without, by some miracle, being killed.[14] A moving but unreliable account has Bill O'Leary leaving Dolly at Glenorchy and travelling up on the *Earnslaw* to visit her periodically, with her always responding to his whistle. One day he returned to the steamer at sailing time in a despondent mood. She hadn't responded to his whistle. He confided in the *Earnslaw* skipper that if Dolly was too old to go back to the hills, so was he.[15]

By this time, however, Dolly was dead. Their partnership had been broken during the incident that nearly ended Bill O'Leary's life. This was the time he tumbled over the 70-foot cliff and was badly hurt. According to one account, before he reached Lake Wakatipu, Dolly stumbled and broke a leg. Another has her falling into a hole at Kinlock on the shores of the lake, with no way of getting her out.

The outcome was the same. In what was described as one of the hardest tasks he ever had to face, and one of the saddest happenings of his life, he

"His mount stumbled, breaking a leg."

An artist's impression of the end of the 20-year partnership between Bill O'Leary and Dolly.

had to "shoot the poor beast" before continuing on to Dunedin Hospital.[16]

Bill O'Leary had bought Dolly as a three-year-old at Waipara in the Arawhata Valley for £5. They were together 20 years,[17] although her age has grown with her master's legend. Tales are told of her living until she was at least 30 and up to 40 years of age.[18] Bill O'Leary described her as an excellent pack-horse. Even though she had never been properly broken in, she was also entirely satisfactory to ride. She would never let him fall off and carried him over many streams in her day.[19] Some said he loved her as if she were his own child.[20]

Bill O'Leary had promised the young Elfin Shaw that he would leave her his horse, saddle, bridle and boots. By the time he departed the district, Dolly was dead, his boots went with him, someone stole the saddle and only the bridle was left for her at Elfin Bay. It was very brittle. It hadn't seen oil for years. She didn't mind. She treasured it because it was his.[21]

Trips into the back country might have been over for Bill O'Leary, but his wandering days were far from behind him. In the winter of 1940, a letter from Wellington arrived for him at the Thompson farm. To his great surprise it was from his youngest sister Clare, now Clare Somes. She had been just as surprised when flicking through the latest copy of the *Free Lance*. Inside was a photo and a write-up about "William O'Leary, the famous 'Arawata Bill'", who for 45 years had explored and prospected for gold in the rugged back country beyond the head of Lake Wakatipu and was living in retirement in Queenstown.[22]

For Clare Somes it was like reading a message from the grave. She thought her brother had been dead for about 30 years. The last she had heard of him was when she was living in Dunedin and he had been missing in a snowstorm for about three weeks. They said there was no possibility of him being alive. As soon as conditions were favourable the police were going out to bury him. She had then moved to Wellington and never heard that he had been found alive.[23] The story doesn't neatly fit either the incident involving Tip Cashmore in the Pyke, with which it has been linked,[24] or the one involving Jim Nolan in the Arawhata.

Clare Somes immediately invited her brother to visit her and maybe take in the 1940 New Zealand Centennial Exhibition at the same time. She told him he would also get to see his youngest brother Jack, who had also settled in Wellington.[25]

With the exception of a doubtful trip to Australia in his youth, Bill O'Leary had never been further afield than Greymouth. This did not deter him from replying that he was coming to see them. The 74-year-old was possibly unaware of the distance, however, because he initially declared that he intended walking from Queenstown to Wellington via the West Coast, before opting for a cheap, second class, crowded boat trip.[26] He bought a new suit for the occasion. The three-piece mail order suit was far too big for him. Elsie Thompson, a trained seamstress, was appalled and wanted to take it in, but he insisted that wear and rain would knock it into shape.[27] He made one further concession and had his full white whiskers trimmed.

In the Wellington suburb of Miramar, Clare Somes prepared her family for the visit of their long-lost relative. She warned them that he had lived all his life in the bush. He would have forgotten the manners he had been taught at home. He would have lost any finesse or polish. He might not be dressed very well. They were not to laugh at him, however, because he was her brother after all.[28]

Her fears were largely unwarranted. He turned up in his nice new suit, collar and tie, hat and polished boots. He always picked a flower for his button-hole and never failed to get to his feet when a lady entered the room. There was one source of annoyance. The first thing he said to her was that he thought she was dead and buried long ago. She reminded him that he was the second-oldest in the family and she was the youngest.[29]

When it came to church-going, brother and sister were each apprehensive about the other, and again it was largely unwarranted. On the first Saturday night of his visit a fidgety guest asked his sister what she did on Sundays. He was relieved to hear that she went to church. He was taken aback when she told him she didn't suppose he had seen the inside of a church for years because their brother Jack didn't go from one year to the next. He told her he walked two miles to Mass every Sunday, and passed around the plate.[30]

At church together the next morning it was Clare's turn to worry. Her brother put a hand in a pocket and brought out a flour bag. After much undoing, he finally revealed his prayer book which he wrapped inside to keep it clean. When he put half a crown in the collection plate, for a moment his sister thought he was going to take some change.[31]

Bill O'Leary didn't much care for Wellington, but there were some things to which he took a shine. He was intrigued by the city's trams, the flash Fiducias, although he wondered why they didn't wear out running up and down the way they did all the time. He was also fascinated by the electric lights and the wireless in the Somes household, which he listened to for hours.[32]

Above all, he enjoyed the Centennial Exhibition, which he attended with brother Jack. Bill O'Leary later wrote to a friend that he had a fine time at the Exhibition, which was well worth seeing, and only cost £18 for the entire trip, including the return boat fares.[33] He loved Playland in particular, going for rides on the rollercoaster, the very high Jack and Jill slide and the scenic railway,[34] again belying his fear of heights. He had already gained a taste for the fun of the fair while attending the New Zealand and South Seas International Exhibition in Dunedin in 1925. This was in spite of a slight mishap coming down the big corrugated slide on a mat. He tried to reduce his speed by using his leg as a brake and ended up completing the ride in an undignified head-first.[35]

The Wellington reunion was special. The O'Learys were not a close-knit family. Timothy O'Leary had begun the scattered family tradition by clearing out to the West Coast, where he died of a heart attack on the Hokitika riverbed in 1900.[36] Mary O'Leary had gone mad and spent the last decade of her hard life in a couple of lunatic asylums. In 1898 she had been locked up in the Seacliff Asylum near Dunedin. Two years later she had been transferred to the Porirua Asylum near Wellington, to be closer to her children Jack and Clare. The 69-year-old had died in the asylum in 1907, and was buried in the nearby Porirua Cemetery. The hospital's medical superintendent said she looked nearly 80 at the time of her death.[37]

The eight O'Leary children had gone their separate ways. This was the first time Bill O'Leary had seen his brother Jack and sister Clare for several

Long-lost brothers Jack (left) and Bill O'Leary visiting the New Zealand Centennial Exhibition in Wellington, 1940.

decades. He also met a member of the next generation, Clare's teenage daughter Stella. His oldest sister Ellen had married and settled in Fiji. Middle sister Mary had married and gone to live in Malaya.[38] No one in the family seemed to know what had happened to brother Frank, although he possibly died at an early age.[39]

Like Bill, the other two brothers, Joe and Dan, had followed their father into gold mining. Both had worked claims in the Queenstown district in the 1880s and 1890s.[40] Bill was able to update his sister on the fate of Dan. He was over the West Coast getting drunk every night of the week.[41]

If there was nothing mysterious about Dan's demise, there was about the fate of Joe. It's known that he mined at Cardrona in Central Otago and did a bit of shepherding and worked as a tourist guide around Glenorchy. Bill O'Leary was either supposed to have gone to the Glenorchy area with his big brother, or followed him there, and they knocked about together for a while.[42]

Joe O'Leary also did some climbing in the area, and had a similar experience to his brother with geographical misnomers. In 1893, Joe and a companion climbed Mount Earnslaw at the head of Lake Wakatipu. In the newspaper report of the climb, which appeared under their names, he was referred to as Joseph Leary.[43] The following year, he guided a party to the top of Mount Earnslaw, the climb being notable as the first ascent of the mountain by women. On their way to the summit, they climbed to the top of a peak which, in a letter written at the time about the climb, he referred to as O'Leary's Peak,[44] while a contemporary newspaper report of the climb called him Joseph ended up being officially known as Leary Peak, until the New Zealand Geographic Board in 2003 finally corrected the name to O'Leary Peak, and Joseph ended up being commonly referred to as Joseph or Joe Leary.[46] According to one explanation, Joe had become a Presbyterian and dropped the filial O,[47] although this is not part of the family folklore.[48]

Later on, Joe O'Leary was employed on a gold dredge on the Kawarau River which flows out of Lake Wakatipu. He was reported to have fallen overboard and was presumed drowned. They never did find the body. There were suggestions that it was no accident. Joe was the only one of the five O'Leary boys to have married, but what was described as a rather unfortunate marriage ended in the then shocking scandal of divorce. Joe was said to have been very depressed. The suspicion was that he had committed suicide. It was also suggested that he wanted to disappear and the not uncommon loss of crew members from gold dredges provided an easy way out.[49]

The staged death scenario is given credence by a remarkable coincidence involving the middle sister, Mary. She was travelling through Australia on her way from Malaya to New Zealand for a holiday. During a stop on the train trip to Sydney she bought a magazine. On the front page was a photo of a J.T. O'Leary, who had just died. The initials were the same as those of her brother, Joseph Timothy O'Leary, and there was a strong family resemblance. Not only that, but the magazine story told of his being born in Lawrence in Otago the year her brother Joe was born, and emigrating to Australia the year her brother was supposed to have drowned. He had worked at a number of jobs, bought a bit of land, added to it and died a wealthy sheep station owner. He was survived by a wife and family.[50]

CHAPTER TWELVE

GLENORCHY
& GOLD

*I*t wasn't long before Bill O'Leary became weary of big city life in Wellington. He went back to Queenstown and then attempted a return to the West Coast. It was a move for which Dan Greaney later took responsibility. Dan Greaney worked as a roadman on the eight-mile stretch between Jackson Bay and the Arawhata River, but after Bill O'Leary had left South Westland. The two didn't meet until Jock Edgar pointed out "the old chap" to Dan Greaney in Queenstown at New Year in 1940:

> So I went over, introduced myself and told old Bill that I was living in the Arawata River and you should have seen his eyes light up. And he really got cracking then. He started talking about the Arawata and waving his arms around and before I left him he said, "I'm coming over. I'm coming over."[1]

When Dan Greaney returned to Queenstown about a month later, the local police constable, Pat Dougherty, said he was worried. Bill O'Leary had gone into training. For the previous week or two he had been tearing up and down a steep hill called Bobs Peak at the back of Queenstown. Pat Dougherty was concerned that if the old man did head for the Coast, he wasn't going to get very far. Dan Greaney later heard that he got as far as Glenorchy after his period of training, but the police intervened and made him return to Queenstown.[2]

After a stint living in a small hut near the Frankton Cemetery,[3] Bill O'Leary ended up back in Glenorchy. He went to stay with Stan and Kath Knowles at the large wooden two-storeyed Mount Earnslaw Hotel. The background to the move illustrates the extent to which Bill O'Leary's nickname had effectively replaced his proper name.

Stan Knowles said to one of the deck-hands on the *Earnslaw*, "If you see old 'Arawata Bill' about, will you tell him there is a job here for a month or two." Stan Knowles was mystified to receive a brief telegram from Queenstown a few days later. It read, "Arriving Friday, O'Leary". The publican couldn't recall anyone of that name and thought it must be a tourist making a booking. When the person duly arrived, Stan Knowles was surprised to learn that O'Leary was Arawata Bill's surname. Bill O'Leary was equally astonished to learn that the man who had sent for him didn't know his surname, even though they had known one another for years.[4]

In his younger days Bill O'Leary would have ridden or walked the 30-mile bridle track round the side of Lake Wakatipu to Glenorchy. On the way he might have stopped at Mount Creighton Station, where he lived for a time in a hut and where he left his last camp oven.[5] These days he was happy to make the trip in style on the steamer.

At Glenorchy he was given the usual job of cowboy and rouseabout at

The Mount Earnslaw Hotel at Glenorchy. The hotel was later burnt to the ground.

the town's only remaining hotel. People began to notice that he was developing the habit of talking away to himself and the cows while he herded them into the byre and milked them. He also fed the hens, helped out in the garden and prepared the vegetables for the hotel meals. When his chores were completed he would often be seen sitting in the sunshine on the north-facing verandah, studying the *Otago Daily Times*. He read it as though it was a book, from the classified advertisements of the front page to the classified ads on the back.[6]

The weather and stories from the newspaper about the progress of the Second World War were always good starters for a chat with one of the locals or hotel guests, who included an increasing number of scheelite miners. The British loss of tungsten supplies in the wake of the Japanese advances into the Pacific had resulted in an influx of miners to work the local deposits of scheelite from which tungsten was extracted to toughen steel.[7]

The scheelite men were more interested in talking to the prospector about mining and minerals than the weather and the latest news. They did not get very far, as the newly-arrived mines surveyor and later Lake County Council chairman and Mount Earnslaw Station owner Tommy Thomson soon dis-

C. DE VRIES

Bill O'Leary stopping for a photograph at the head of Lake Wakatipu.

covered. He was eager to glean all he could about the geology of the area. The pub cowboy was more interested in talking about sea-boots full of gold in West Coast caves.[8]

Bill O'Leary was his usual big hit with the children at Glenorchy. Vic Yarker, who had first met him while working on the Milford Road in 1935, was staying at the pub with his wife Ailsa and young family while helping to build accommodation for the scheelite miners. Ailsa Yarker recalled of her fellow guest:

> He would sit on a little stool out the back of the hotel and Evelyn, our young daughter who was about five, would come home from school and she would help him do the carrots or whatever vegetables he was preparing for the hotel. I thought he was a dear old soul and so did Evelyn. She thought he was lovely. All the kids that lived about there thought he was marvellous and they loved to talk to him.[9]

Bill O'Leary's best friend at Glenorchy was Matt Hunter. They were about the same age and the Knowles were looking after both of them. Matt Hunter was a painter by trade who did odd jobs around the pub. The two were chalk and cheese. One the quiet non-smoking near wowser with snow white beard. The other the singing piano-playing boozer whose moustache had turned ginger as a result of the smoke from the cigarettes hanging from the corner of his mouth. He could down a gallon or more of beer in an evening and the more shickered he became the better he played. At night they went their separate ways. Matt Hunter would be in the bar and Bill O'Leary would retire to the pub sitting room or to bed in his hut out the back.[10]

By this time, Bill O'Leary's mining days were well and truly over, but he was still renewing his miner's rights every year. This he had done diligently throughout his mining life, either through the Hokitika or Queenstown wardens' courts.

The contents of his wallet which was found at Deadmans hut illustrates

Entries in the Queenstown Warden's Court register of miners' rights 1933-42.

how conscientious he was, while the dates and places of issue reinforce the
wandering nature of his life. As well as the note from Charlie Haines, the
wallet contained a wad of five shilling miner's rights covering the years 1927
to 1931. They were issued at Okarito in May 1927, Glenorchy in May 1928,
Hokitika in June 1928, Glenorchy again in May 1929, back to Hokitika in
June 1929, Glenorchy in October 1930 and again in October 1931. For two
of those years, he had miner's rights issued at both Glenorchy and Hokitika.
One was enough to cover the whole of New Zealand.[11]

During the 1930s Bill O'Leary continued to turn up each year at Glenorchy or Queenstown to renew his miner's rights. As late as November 1941
the 76-year-old fronted up to the postmaster at Glenorchy to have a miner's
right issued for a further 12 months.[12]

About the same time as the Deadmans hut wallet was discovered, Murray
Gunn recovered from the hut Bill O'Leary's pick minus its handle, his watch
case minus its internal workings and chain, and his wooden hairbrush with

The Arawata Bill display case (left) in the Hollyford Museum before the building was burnt down.

'ARAWATA' carved on the back. These items, along with the wallet of miner's rights and note from Charlie Haines, were put on display in Murray Gunn's Hollyford Museum at the former Public Works Department camp in the Hollyford Valley. All but the pick were destroyed in a fire which gutted the museum in 1990. The pick and other iron work was salvaged from the ashes and put on display in a replacement museum built soon after.[13]

For all his years spent looking for gold and other precious metals, Bill O'Leary never found much. A flash of gold here, a speck there, a few small nuggets, a few mineral samples, nothing to keep secret or write home about. Sometimes he made just enough to keep him going, other times not enough to make ends meet without picking up casual work to subsidise his prospecting.

It's hardly surprising. He was generally regarded as more of a wanderer and a fossicker than a hard-core miner. They said he was too erratic, not methodical enough. He never stayed in one place long enough to find anything of real value, never settled down to mine one claim. He always poked around

Bill O'Leary's miner's pick at the Hollyford Museum.

near the surface, never went underground. Most of the places he searched were either not gold-bearing country or contained little or no payable gold. There were also doubts about his knowledge of minerals and mining methods. Above all he lacked that essential element, luck.[14]

In the case of all the time he spent in the Arawhata Valley, a further element has been offered. He was the victim of a practical joke. At least that's the story Jack Cron senior heard and passed on to his sons. The story was that Bill O'Leary and two other men went into the Arawhata from the Otago side to do some prospecting. One of them had in his pocket a nugget he'd brought from the Shotover River. He took the nugget from pocket to dish and pretended he'd just panned it. The joke fell flat when he confessed and Bill O'Leary refused to believe his mate was having a bit of fun.[15] It's a good story and probably a bit of fun in its own right.

Despite his lack of success, Bill O'Leary remained eternally optimistic, as he demonstrated in newspaper and magazine interviews in later life. He conceded that he had never discovered gold "in anything like quantity" during his long years of prospecting, but remained convinced there was a good deal of gold to be won in some parts. Large deposits lay hidden in the mountains and dredging would yield fortunes from the rivers, particularly the Arawhata. Most of its bed was of crushed quartz containing fine gold, and he was picking that some day they would be dredging the 18 miles or so which were suitable. There were also rich quantities of gold to be had at Bruce Bay, where the sea seemed to throw it up about every 10 years.[16]

It was a theme he continued in conversations with acquaintances. He told a government survey party member at Waiatoto, Jim Bourke, about a big terrace in the Arawhata where he would make his fortune if he could get water onto it.[17] This would explain his tremendous efforts to haul discarded fire hoses through the Ten Hour Gorge.

According to Martins Bay miner Edward Martin, if any of them found anything, he would tell them he was following it up in his lead. He had leads everywhere.[18] One such "crater" lead he claimed to have spent five years tracing for more than 100 miles across the rugged shoulders of the Skippers and Olivine ranges.[19] Similarly, Davie Gunn told of his friend being driven by:

> ...a junction between schist and granite rock which Bill thought he had discovered, a junction which could be seen, right enough, at various widely scattered places throughout the countryside, but which existed more largely in his own imagination. He dreamed of it as a marvellous discovery which would be equal in importance to the Rand in South Africa. "My lead", he called it, and although there was really little or nothing in it, it no doubt made him in his thoughts a wealthy man.[20]

It was a powerful incentive. As he told Jack Cron junior, "If you can get a little bit of gold Jack, you'll make enough in a week to do you a lifetime."[21] He never did.

In addition to a bit of alluvial gold, Bill O'Leary told a correspondent from the *Free Lance* in 1939 that he had found traces of practically every kind of mineral, including cinnabar, platinum, rubies, manganese, white sapphires, oil and copper, although mostly they had been found in only very small quantities. The *Free Lance* report continued:

> At those picturesquely-named places Lake Alabaster and Madagascar Beach there is shale which will burn from the lighting of a match. The two hot springs, Waiho and Copeland, are on a lode which runs right through the country, "Arawata Bill" says, and it is a valuable one full of minerals. It varies in type according to the country through which it passes. Sometimes it is granite and then rubies or garnets are found. If it is of schist or shale, it bears quartz-reef; if it is of glass rock mica is probably there. There is another load in Red Hills which contains a body of iron three feet wide. There is also asbestos between the Martha and Cascade rivers. Indeed, the old prospector declares there is no shortage of asbestos or iron in the district.[22]

There was coal too, although he did not know whether it would be payable to work coal deposits so far from transport. It was the same story with valuable finest quality native timber of which there was enough to fix the housing shortage if the transport problem could be solved.[23]

Bill O'Leary was also full of praise for the land itself, particularly the rich alluvial land around the Arawhata and Waiatoto Rivers, which he described as grand cattle country. He predicted that some day it would be opened up and would add greatly to the wealth of Otago. All it needed was a road to be put through.[24] With this in mind, he added his voice to a 1935 Otago Expansion League call for a road to be completed over the Haast Pass. He claimed that the Haast River Valley, which could be tapped from the road, had a depth of arable soil of 15 feet, even more than the Wairau Valley in Marlborough.[25] Whereas cattle on the river flats were regarded as an asset, he was concerned that they were destroying the undergrowth, and that young trees, which were a source of future timber, were becoming scarce. Bill O'Leary looked on deer as a nuisance for the same reason. He complained that herds of 30 or more Virginian, red and fallow deer caused great damage to the bush, and there were numerous places where the undergrowth had been completely wiped out.[26]

There is some evidence to support Bill O'Leary's discovery claims. He was known to carry bottles of shale oil and a few samples of rare minerals which he had analysed during annual trips to Dunedin.[27] The minerals were carried in sample bags which he bequeathed to Gordon Speden.[28] Furthermore, he once showed Glenorchy resident Wattie Clingin "a wonderful bit of asbestos" with a long fibre about two inches in length, although Wattie Clingin doubted its commercial value.[29]

His reputation was such that geologists were also interested in what he was up to, but without much reward. In 1937, for example, Jas Healy of the New Zealand Geological Survey and geologist Frank Turner were to go with him to look at mineral deposits which he regarded as of value. The trip fell through because Bill O'Leary never turned up to make the arrangements.[30]

Hard evidence or not, there was always the hint of things to come. Neil Cook junior recalled him arriving at the Route Burn after a prospecting expedition. "'Plenty of gold in there,' he said. 'Plenty of gold. Plenty of rubies. I didn't have time to bring any out this time, but I will next time.'"[31]

Sometimes such claims might have been tinged with an element of leg-pulling. Just as a practical joke gone wrong is said to have been behind his search for gold in the Arawhata, it's been suggested he wasn't above playing a joke on others. Glenorchy resident Gilbert Koch recalled the occasion when Bill O'Leary kidded Stan Knowles that he had a plant of rubies in the hills and was going to bring some over for him:

He used to come up to here laughing like a jackass. Dad would say to him, "My gosh what's wrong Arawata?" "Oh," he said, "I was just thinking about Stan down there. Stan Knowles," he said. "He's lapping it up about these rubies," he said.[32]

G.W. TRICKER

One of six Gary Tricker etchings that were hand-printed by the artist from copper plates to illustrate the 1981 limited edition of Arawata Bill.

Red Mountain from the historic Red Hills trig in the Red Hills Range.

The vertical bluffs from O'Leary Pass on the skyline (centre) to Victor Creek and the Joe River below.

The Joe Valley to the right of O'Leary Pass.

R.F. ENTWISTLE

Lake Alabaster. The remains of a Davie Gunn fence enters the lake on the left.

R.F. ENTWISTLE

The Barrier Gorge through which Bill O'Leary blazed a trail from the Pyke River below.

D.G. BISHOP

Lake Wilmot to the east of which the lost ruby mine was said to have been located.

R.F. ENTWISTLE

Bill O'Leary's summertime base of Big Bay.

Learys pass. fr Arawata upper flats

Charlie Douglas' 1884 sketch of the feature he later named Learys Pass

The Arawhata River rages through Gerhard Mueller's Ten Hour Gorge.

I.M. DOUGHERTY

Lloyd Veint's 1980 painting of Bill O'Leary and Dolly.

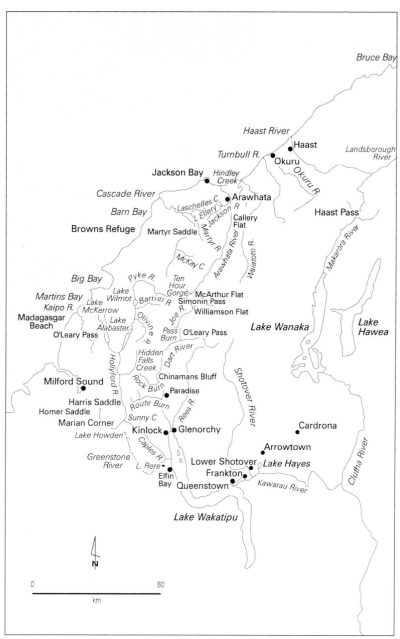

Bill O'Leary's North-West Otago and South Westland.

CHAPTER THIRTEEN

THE LITTLE
SISTERS

*I*n the winter of 1943, Bill O'Leary's health deteriorated. Stan and Kath Knowles were worried about him, particularly because they were about to vacate the pub. They arranged for him to go to a Catholic old people's home in Dunedin. On 26 August he entered the Sacred Heart Home of the Little Sisters of the Poor in the suburb of Andersons Bay. The French-based order of nuns recorded his occupation as "mineur d'or", his marital status as "celibataire" and his next of kin as "sa souer Mrs Somes".[1]

According to one account, "He was delighted to be there and expressed the wish that he had gone a few years earlier to enjoy the comfort of a clean, cosy bed, well-cooked meals, and the companionship of other men."[2] Those companions included his old friend Dan McKenzie. He had decided early

The Little Sisters of the Poor Sacred Heart Home in Dunedin's Highcliff Road. It was demolished in 1979, after a replacement was built in the Dunedin suburb of Brockville.

on that trying to earn a living raising cattle at Martins Bay was a mug's game. He left his brothers Malcolm and Hughie to it, and went to work as a track-man, a gold miner at the Shotover River and a scheelite miner at Glenorchy.

When Bill O'Leary recovered from his winter ills, the slight but still very straight old man with the wiry physique[3] set to work to show his appreciation of the care and kindness of the Sisters. Dressed in white shirt and tie, three-piece suit and heavy coat, he would head outside to tend the vegetable plot[4] and take care of the pigs in the pigsty. Dan McKenzie told a visitor during a guided tour, "You know he's got it that nice and tidy I wouldn't mind sleeping there."[5]

He also continued his habit of reading,[6] and a newspaper report referred to him as a raconteur who was seldom without a circle of listeners.[7] As with many aspects of Bill O'Leary's life, of his talents as a talker there are conflicting memories.

Among some, he had a reputation for skill and authority. He was an interesting man to talk to and a great conversationalist. His speech sounded quite educated. He was deeply read and well versed. He could talk about almost any subject in the world. He had an extensive, exceptional, great, remarkable, fascinating and phenomenal knowledge of the West Coast and

Bill O'Leary (left) and Dan McKenzie in their "declining years" at the Sacred Heart Home in Dunedin.

Fiordland, of gold and mining and minerals, of timbers and bushcraft and river crossings, of history and the Empire. He seemed to be up to date on everything even though he spent many months in the back country,[8] and on it went.

There were others who held the opposite view. He was not what you would call a conversationalist. You couldn't talk to him for any length of time or keep him to the one subject. He would flit from one topic to another, or break in with something completely different he was thinking about. It was sometimes difficult to understand him because he had no teeth and the words were fouled up with his whiskers on the way out.[9] The extreme expression of the view came from Barbara Heffernan:

He used to sit for hours didn't he. ...And he would just sit there. "Well, sometimes I sits and thinks, and sometimes I just sits." You know, he just looked to me as if he was just sitting.[10]

Bill O'Leary was something of a celebrity at the Sacred Heart Home. His visitors included *Weekly News* photographer Bill Beattie, who had photographed him at Big Bay 17 years earlier.[11] He also met up with climber George Moir of *Moir's Guide Book* fame, who went to the Home to visit Dan McKenzie.[12]

Other visitors included the *Wide World Magazine's* Dunedin correspondent Harry Fortune, who had written the second-hand article about him which appeared in 1935. This time Harry Fortune decided to get some first-hand information in preparation for a further article for the London-based magazine. Fortune later reported that a copy of the original article had found its way into the old man's hands and he had been so thrilled that he had carried it with him all the time. At the Sacred Heart Home the article was still his proudest possession, even though the pages were so thumb-worn that he showed them to very few people. "They even know of me in London," he told the correspondent.[13]

Bill O'Leary was no doubt flattered by the publicity, but also quietly modest, a bit bemused and a shade embarrassed. As he told Harry Fortune, "I don't want you to write anything sensational about me. What I've done any man could do."[14] Bill O'Leary added that he would like to see the story, although he didn't think it was a good idea for a man to be made a fuss of.[15]

It was a sentiment he had earlier expressed to Dan Greaney during their meeting in Queenstown:

> I asked him quite a few questions about his life and he said to me, "I'm a very much over-estimated man." I thought that was the under-statement of the year knowing some of his exploits.[16]

Similarly, an *Otago Daily Times* reporter noted that the Sacred Heart Home resident was not fond of publicity. It was natural, however, that a man who had led such an unusual life should have become well known and he was proud to have figured in a magazine published in London.[17]

He presumably had the same mixed feelings towards the appearance in 1947 of Alice McKenzie's memoirs *Pioneers Of Martins Bay*. In the book she devoted a couple of pages to the one-time Martins Bay prospector, including the accounts of his running away from home and the incident involving Cashmore and the corpse.[18]

To the *Otago Daily Times* reporter, Bill O'Leary seemed settled, or at least resigned to life in the Sacred Heart Home:

> In the distance beyond the hazy city the hills shone clearly in the morning sun. The old man's keen blue eyes sought them out as he said a little wistfully, "No, I am past feeling the urge to go wandering. You can't do it at my age. It is all in the past and I am happy here with my gardening, my reading – yes, and with my memories."

The reporter added that it seemed the call of the unknown, of bush tracks and of quiet hills was still with Bill O'Leary, even if he could no longer answer it.[19]

Similarly, he told Harry Fortune that he was very happy in the Home:

"But" – his eyes turned wistfully toward Dunedin's hills, and I knew he was thinking of the remote region he had made his own – "But out there..." I saw tears in his eyes, and realized something of the tragedy of old age when the spirit remains eternally young.[20]

Tom Kennett, who had moved from Glenorchy to work in Dunedin, was less sentimental, less lyrical, more blunt. On the basis of his visit to the Home, he wasn't so sure that his old friend was happy with the order and routine of institutional life in the big city, after decades of outback tent and hut. The Little Sisters were good to him and he appreciated their care, but he didn't like the confines of some strange environment which didn't suit him.[21]

Bill O'Leary soon demonstrated that he was not quite past feeling the urge to go wandering. One morning in 1947 in the Wellington suburb of Miramar, his sister Clare went to collect the milk at seven o'clock and he was sitting on the doorstep. "You get up late," he told her. When she asked him where in the world he'd come from, he simply replied, "Off the boat, of course." The 81-year-old was in bad shape. Travelling alone to Wellington, he had been forced to take a shakedown on the crowded ship and had been robbed while he slept on the floor.[22]

Youngest brother Jack had died since Bill O'Leary's 1940 visit to Wellington, as had his brother Dan on the West Coast. Instead there was a third generation to greet him. His niece Stella was now married with a two-year-old daughter, Shonah. He adored his grand-niece. He loved taking her for walks and buying her pottles of strawberries. When his sister protested that a whole pottle would make her granddaughter sick, the next time they went walking he let the little girl finish the pottle before they arrived home.[23]

As always, his manners were to the fore. After his niece Stella had taken him to the shipping company to arrange his return berth, they travelled home by tram. Stella Dalley later recalled how there was only one seat left in the men's compartment:

> I wanted him to sit down because I was used to standing in trams and he wasn't. But he looked round all these men and said, "I will have you know I was never brought up to sit while a lady stands," and straight away four or five of them got up and offered me their seats. So I said to my mother, "He might have lived in the bush all his life but he showed up a few city chaps on the tram."[24]

During this second visit to Wellington, Bill O'Leary became the subject of another police search. It had its origins in his West Coast and Lake Wakatipu induced habit of greeting vessels as lifelines.

His sister Clare did not live far from the oil tanker berth at the Miramar Wharf. The huge tankers fascinated him. One Saturday morning he took off before breakfast to farewell a departing tanker. He didn't return for breakfast or lunch. When tea-time came and a family search had failed to find any sign of him, the police were notified.

About half-past nine they located him wending his way home the last few hundred yards. His sister told him she was worried that if he was out all night, he'd get pneumonia and that would be the finish of him. He told her she needn't worry. He'd slept out all night in the snow and never got pneumonia before.[25]

When they asked him where he had gone after he left the wharf and whether he had got hungry, his niece recalled a simple explanation:

> He said, "I went up that road that goes up to the top of the hill. There's a huge boarding-house up there and I told them who I was and where I'd come from and they took me in and gave me some tea and sandwiches." Poor Bill, he did not know he had got into Mount Crawford Prison.

From there he had walked into the city centre. He didn't believe the police had been out looking for him. The policeman he'd been talking to in Courtenay Place didn't mention it.[26]

After about three weeks in Wellington, the aged traveller was anxious to get back south.[27] He returned to the Sacred Heart Home, but still couldn't settle. By late June 1947 he was again heading for the hills.

It was a move for which Dan Greaney also took responsibility. He wrote to Bill O'Leary at the Sacred Heart Home asking for information about the last time he had seen kakapo in the Jackson Bay area:

> I enclosed a copy of my description of a dawn chorus heard at Lake Ellery in about 1943, a wonderful chorus of many thousands of bellbirds and tuis singing at dawn. Also I sent an enlargement of the Arawata Valley and River, about the worst thing I could have done, although I meant well.

A week or two later Dan Greaney heard that Bill O'Leary had absconded.[28]

According to contemporary accounts, a combination of factors enticed Bill O'Leary away from the snug winter comforts of the Home. He wanted to visit some old friends. Still affected by prospecting fever, he wanted to recover the cherished gold-panning dish he had stowed away in the area and prove to everyone in the Home that he could still find gold. Above all, he was answering "that call which would never be stilled – the call of the wild." Every now and again since he had gone to live in Dunedin, the reports continued, he had felt periodic urges to return to the back country and resume his wandering life. The most recent impulse was stronger than usual. The Little Sisters of the Poor had tried to dissuade him, but he was adamant.[29]

Bill O'Leary caught a bus for Queenstown, but he had barely recovered his panning dish and begun prospecting than the weather changed. Caught out in the open in a mid-winter snow storm, he was forced to sleep under a tree in the freezing cold with thick snow on the ground at Bendemeer Station near Lake Hayes.[30] The next morning he turned up at the Lower Shotover Pub before breakfast with a billy in hand and asked the publican, Vic Gerken, for water. He was still there in the early evening when Albie Collins and a mustering mate bumped into him. Albie Collins had come to know him while delivering meat and bread orders to his hut at Frankton. He renewed his acquaintance with Bill O'Leary, who was sitting with his back to a roaring fire.[31]

"Weren't you cold?" Albie Collins asked him. "I would have been if I hadn't had this," he replied, gesturing to his navy blue overcoat. Albie Collins reckoned he must have been "the toughest rooster that ever walked". Even with the coat, anyone else would have frozen to death sleeping out that night. Bill O'Leary later confessed to Harry Fortune, "I couldn't stand the cold. And it was cold." While Bill O'Leary and Albie Collins were chatting, Vic Gerken had a phone call from the Little Sisters of the Poor and the wanderer was returned to Dunedin.[32]

Like so many stories about Bill O'Leary, this one also has several versions. According to the account Dan Greaney heard, someone spotted the traveller getting off the Dunedin bus at Arrowtown and heading west. The local constable rang his opposite number in Queenstown and they agreed to meet half way. They found Bill O'Leary sleeping under a plum tree in about six inches of snow.[33] Variations have him walking all the way from Dunedin and sleeping under hedges as he went,[34] and returning to the Home after discovering that someone had removed most of his mining gear from his old claim.[35]

According to the contemporary reports, the Sisters were amazed that a man of his age could do what he had done in the previous few days and not be in the throes of pneumonia. Many a younger man would have died, they continued, but his early life must have stood him in good stead. His tough old hide and magnificent constitution did not fail him and after a few days in bed at the Sacred Heart Home he was none the worse for his ordeal. He told one of the Sisters it was a good bed and better that lying out under a tree.[36]

The contemporary reports observed that for the moment, Bill O'Leary was convinced his prospecting days were over. The Sisters certainly hoped it would be the last of the old man's wanderings and he would settle down to life in civilisation. One report qualified this by adding that he would still no doubt cast a nostalgic eye over his shoulder to the hills of Central Otago and the country in which he spent more than half a century.[37]

He did more than that. Four months later a frail Bill O'Leary wandered away from the Home in a second bid to return to the country he loved. Tom Kennett saw him shuffling up a hill suburb on the southern route out of Dunedin:

> I was working on the roadside out of Dunedin there up near Mornington and someone says to me, "There's an old codger getting along the road." I looked over and I said, "By Jove, I know that fellow, it's Arawata Bill." And I went over and talked to him and he says, "Oh," he says, "I'm heading for Queenstown." He had nothing but a pair of slippers in his coat, sticking out of his pocket, and I says, "Oh well you've got a long walk haven't you?" "Oh," he says "I want to get to Queenstown," he says. "I've got some stuff over in Arawata that I want to have a look at too."[38]

Tom Kennett had no doubts about his friend's mental state. "He was quite sensible. He knew all we were talking about."[39] Others were not so sure. According to Dan McKenzie, "Old 'Arawata Bill', like so many men who spent years in solitude, went quite 'wampy' in the end."[40] Tom Kennett was worried enough about his friend's physical state to get a message to the Little Sisters of the Poor and they came and picked him up.[41]

Bill O'Leary was too ill to be returned to the Sacred Heart Home. In the words of a local newspaper, his strength was not equal to his spirit, and he was taken to the Dunedin Public Hospital.[42] There, a week later, on Saturday 8 November 1947, William James O'Leary died, 11 days after celebrating his 82nd birthday. The doctor who examined him the previous day listed as the causes of death, "Ventricular fibrillation, Uraemia – 7 days, Prostatic obstruction – years, Myocardial degeneration – years."[43] In short, cardiac arrest as a result of kidney failure.

At seven o'clock on the Tuesday morning, a Catholic priest, Father Cyril Ardagh, conducted a requiem mass for Bill O'Leary at the Sacred Heart Home. A private interment followed in an unmarked grave purchased by the Little Sisters of the Poor at the nearby Andersons Bay Cemetery.[44]

Bill O'Leary's death was marked by prominent and lengthy obituaries in several daily newspapers. Tributes were paid in Dunedin's *Otago Daily Times*[45] and *Evening Star*, which described him as one of the most colourful identities of South-West Otago.[46]

Obituaries also appeared in Invercargill's *Southland Times*[47] and *Southland Daily News*. It referred to him as a man of nature, an explorer, a hardy man who cared little for comfort and knew little of civilization, a romantic man driven always by dreams of wealth, as happy with his own company as with that of his fellow men, a man of the bush.[48]

Wellington's *Evening Post* lamented that with his death, the West Coast and New Zealand as a whole bid an affectionate farewell to one of the most picturesque of its old identities. It added that while O'Leary Pass on the Barrier Range perpetuated his name, even more substantial were the affectionate memories of all who knew him, and the respect and admiration of the new generation who followed in his trail.[49]

The Auckland-based *Weekly News* and Wellington's *Free Lance* also paid their respects. The *Weekly News* referred to him as a famous and rugged character, and the most colourful figure in Otago.[50] The *Free Lance* described him as a romantic and colourful personality and a famous hermit, explorer and prospector. It added that his passing was regretted by many who admired and respected those sterling qualities of that fast-departing school, the early pioneer.[51]

Obituaries also appeared in the Canterbury Mountaineering Club journal, the *Canterbury Mountaineer*[52] and in the London-based *Wide World Magazine*.

It appeared courtesy of Harry Fortune, who had interviewed Bill O'Leary for the follow-up article just before he died. It referred to him as a most colourful New Zealand character with no equal in the country's history, and concluded with the traditional lament, "We shall not see his like again".[53]

CHAPTER FOURTEEN

PLOTS
& PLAQUES

*F*or more than a decade, visitors to the Andersons Bay Cemetery could walk past Lot 64, Block 191, oblivious of the individual who was buried beneath the bare plot.[1] It bothered some of those who had known Bill O'Leary, including Jim Reid, the caretaker of the Queenstown Borough Motor Camp and a former resident of Glenorchy. He had attended Bill O'Leary's funeral service in 1947. When he visited the cemetery 10 years later, the grave was still unmarked.[2]

Jim Reid mentioned his concern to the president of the Southland Tramping Club, Peter Chandler. He remarked that many of the well-known characters of the area, such as Dan, Malcolm and Hughie McKenzie, Davie Gunn and Bill O'Leary, were all dead and gradually being forgotten. While none of them became wealthy or laid great claims to fame, they all played their part in providing some colour in the story of the back country of North-West Otago. If some organisation such as the Tramping Club would sponsor an appeal, Jim Reid felt sure quite a few people who knew Bill O'Leary would be prepared to put in a few bob to help.[3]

The Club took up the suggestion. Peter Chandler first approached the Little Sisters of the Poor for their permission. The reply came from the Superior of the Sacred Heart Home, Sister Rosa:

> The Old Gentleman had been some time in our Home, but towards the end of his life he became very erratic and would not stay, having been taken ill in the street he was taken to the Hospital and as he had no relatives we again interested ourselves in him and when he died we made

his funeral arrangements, which is the reason why the plot would be in our name, we have no objection at all to the erection of a stone or anything of that nature.[4]

The no relatives comment only applied to Dunedin. He still had surviving relatives outside the city, including his sister Clare who was given as his next-of-kin in the Sacred Heart Home admission book.[5]

Having secured the Little Sisters' blessing, Peter Chandler went public with letters to the editors of the *Otago Daily Times* and the *Southland Daily News*. He explained that Bill O'Leary lay in an unmarked grave in the Andersons Bay Cemetery and that the Southland Tramping Club was sponsoring an appeal for funds to cover the cost of erecting a simple memorial stone, "so that his memory and explorations shall not be forgotten".[6] The *Otago Daily Times* and the national weekly newspaper *New Zealand Truth*, also ran news stories about the memorial proposal.[7]

The response was overwhelming. The club received 49 donations totalling £34 2s. The donations reflected Bill O'Leary's entire life. One was sent by someone describing themselves as an "Ex Wetherstonsite". Money came in from a couple of old West Coast friends, Frank Heveldt and Dinny Nolan, who also sent a copy of the poem he had presented to Bill O'Leary when he left South Westland. Former Big Bay miner Colin McEwan sent a donation, as did Bill O'Leary's Ayrburn Station workmate Dave Matheson and his former employer, the Westland County Council.[8]

Donations were made by those who had known him at the head of Lake Wakatipu: Donald and Nancy Watherston from Glenorchy, Doug Scott from Rees Valley Station and Emily Forsyth, formerly Emily Hallenstein, who had remembered him from her younger days when he would visit the Aitken family at Paradise. Money was received from climbers Gordon Speden and Arnold Heine, along with tramper Lindsay McCurdy's daughter Mary McCurdy and tour guide Arthur Hamilton. The Otago Tramping Club forwarded a donation. So did George Moir and Bill O'Leary's niece Stella

Bill O'Leary's headstone, which was erected in early 1959. In the top left corner are two miner's picks crossed over a miner's shovel.

Dalley. Jim Reid and Peter Chandler were among the other individuals who made contributions.[9]

As a result of the public appeal, a headstone was erected in early 1959.[10] Peter Chandler was able to write to a friend, "'Arawata Bill' is safely anchored under a large lump of Bluff norite which ought to put a finish to his wandering."[11]

It had been suggested the memorial should be made of stone from "Bill's country" in North-West Otago.[12] That wasn't to be. A Dunedin firm of monumental masons, A.E. Tilleyshort And Co, offered a reduced rate of £17 for the dark grey Bluff norite headstone, which it told Peter Chandler had "two little chips out of one edge, but these are hardly noticeable."[13]

Partly because of the cut price and partly because of the amount donated, less than half the money was needed to buy the headstone. Some of the surplus was used to purchase a small bronze plaque. Given that Dunedin already had a headstone, which many people outside the area would be unlikely to see, it was decided to place the plaque elsewhere.[14] Initially the waterfront wall at Queenstown was considered.[15] The Club finally decided to ask the Lake County Council if it would agree to the plaque's being cemented on the new bridge over the Rees River at the head of Lake Wakatipu. The Club felt this was appropriate because the bridge was the most prominent, permanent structure in the district in which Bill O'Leary spent many years.[16]

The Council gave its permission and the 6x4 inch plaque was duly cemented to a concrete support at the east end of the bridge. It read:

In memory of
William O'Leary
("Arawata Bill")
1865-1947
Roadman — Prospector — Explorer[17]

Even after the headstone and plaque had been paid for, the Southland Tramping Club still had enough money left over for tributes to another southern pioneer, William Henry Homer. He had discovered the Homer Saddle under which the Homer Tunnel was built to enable a road link between Te Anau and Milford Sound. The two men shared a similar end. Henry Homer died in Frankton Hospital in 1894 and was buried in an unmarked grave in the Queenstown Cemetery. The Club decided to mark his grave with a bronze plaque placed on a slab of Hollyford diorite. The balance of the money went to the Fiordland National Park Board to put towards the cost of erecting a memorial for Henry Homer at the entrance to an alpine garden it was establishing at the Milford Sound end of the Homer Tunnel.[18]

For several months motorists travelling over the new Rees River bridge at the head of Lake Wakatipu were able to view the bronze tribute to Bill O'Leary. It then disappeared. Some thought it must have been taken by a souvenir hunter, a memento collector. The local gossip was that a disgruntled and spiteful resident chiselled it off and chucked it into the river. They said he hated Bill O'Leary's guts because he thought the prospector was a scurrilous old rascal, a lazy devil, a useless old bugger. There were a dozen others in the district who were just as much entitled to a plaque.

Some suggested there was an element of sour grapes in the comments. He probably included himself in the dozen. When confronted with the allegation that he had removed the plaque, he denied it.

The erection and removal of the plaque unintentionally acted as a focus for the polarization of views on Bill O'Leary. There were those who shared the criticism. According to them, Bill O'Leary was just an old hermit who was looked on with pity, an old bushy who took likes and dislikes to people without any reason, a contrary cantankerous old fellow, very grumpy, no angel. He was not very alert, not very active, lazy. He never climbed a peak or got up off the flat if he could help it, never worked very hard unless he had to, never had accidents because he never went fast enough.[19]

There were a lot more men, it was argued, who could just as well have been legends. Men who didn't waste their lives wandering around looking for non-existent treasure and then come out during the winter and put their feet under someone else's table. It would have been different if he had found gold or diamonds or rubies, but was a rather pointless sort of life because he never acquired anything, didn't make a contribution, never did any good for the country. Not like the McKenzies or Davie Gunn who tried to bring in the land and breed cattle. Not like Charlie Douglas and those who mapped the place. In the words of one of his detractors, sometimes a man didn't have to do much to get famous.[20]

In the opposing camp were those who were glowing in their praise of the man and his achievements. To them he was a decent character, a real good sort, a great chap, one of nature's gentlemen. He was always in the best of spirits, always had a smile and a laugh for everyone. He was well-liked, a lovable person, everyone's friend. He was an absolutely honest, staunch, straight, generous, reliable bloke who wouldn't take a single penny unless it was owing to him. He was never known to put a crook deal over anybody, never forgot a favour. He was the most colourful character ever met, a rare sort of chap, one out of the box.[21]

Not only was he admired, but many believed he fully deserved the attention he received. To them he was one of the greatest ever back-block bushmen and rivermen, and an excellent mountaineer. He had done so much on his own that nobody else would ever think of doing, such as walking over O'Leary Pass in gumboots. He had earned his fame for the remarkable years he spent in remote areas and the hardships he endured. He was part of the country's history, part of its heritage.[22] It was even suggested that a memorial headstone and plaque were insufficient recognition. He deserved "a statue of inspiring proportions, design, material and alpine location".[23]

CHAPTER FIFTEEN

POETIC
RHAPSODIES

During his lifetime Bill O'Leary had already achieved the status of celebrity, folk hero and legend, at least in Otago, Southland and South Westland, if not further afield. Following his death, that status was extended and enhanced, largely through the work of New Zealand poet Denis Glover. The result was that Bill O'Leary, or rather Arawata Bill, not only became a part of the country's folklore but its literature.

The prospector's tendency to inspire others to break into verse had begun with Coaster Dinny Nolan, who wrote his Arawata Bill tribute in the late 1920s. Twenty-five years later Denis Glover also used the name Arawata Bill as the title and hero of a sequence of 20 poems based albeit loosely on the life of Bill O'Leary.[1]

Chance led Denis Glover to write the *Arawata Bill* sequence. In his Canterbury University student days he had been a keen mountaineer, joining the Canterbury Mountaineering Club in which he was "a very junior member among men who were busy exploring the Southern Alps". They included John Pascoe, and Rod Hewitt, the climber with whom Denis Glover made the first traverse of Canterbury's Mount Rolleston.[2] From these two men came the initial idea for the sequence of poems.

At the beginning of 1953, a bearded John Pascoe shambled into Denis Glover's Pegasus Press workplace in Christchurch. He was on his way home to Wellington after completing a trans-alpine trip from Lake Wanaka to the West Coast with a party of climbers from the Canterbury Mountaineering Club. This was the trip during which they followed Bill O'Leary's trail of stone cairns on the approach to Simonin Pass. John Pascoe told the Christchurch poet of Arawata Bill and the cairns and of finding a shovel that must have belonged to him. Denis Glover wrote at the time that he was "greatly taken with the name 'Arawata Bill', which accorded with my own notion of trying to turn fact into poetic legend, but I did nothing about it."[3]

A few days later an immaculate Rod Hewitt strolled in, fresh back from climbing. When Denis Glover told his old school friend and mountaineering mentor of John Pascoe's magnificent three-week ginger growth on his face, Rod Hewitt informed him that it was his own practice always to shave in the mountains. According to Denis Glover, within an hour he had written an initial poem called Arawata Bill, which included a reference to shaving.[4]

At the time, Denis Glover wrote that previously he had never heard of Arawata Bill. When he retold the story a couple of decades later he was less emphatic, on one occasion writing that he knew the country when he was young and probably even saw Arawata Bill with his cobber Jock Edgar,[5] and on another occasion claiming to have remembered Arawata Bill, although only "hazily".[6]

As a boy, Denis Glover and his younger brother had spent their annual holidays based at a cottage in Queenstown. During one of these holidays the 14-year-old schoolboy Denis Glover had worked for Jock Edgar as a deck-hand on his tourist launches *Thelma* and *Kelvin*. It was one of the loose links in the backgrounds of the prospector and the poet. Denis Glover was from Irish immigrant stock on his father's side. His maternal grandfather started a newspaper in Lawrence about the time Bill O'Leary was born at nearby Wetherstons. Denis Glover was born in Dunedin in 1912, in the city where Bill O'Leary died and was buried.[7]

Denis Glover did not stop at the initial Arawata Bill poem. As he explained at the time:

I had become fired with the idea of Arawata Bill himself, and all the unremembered explorers and prospectors who tackled that terrible corner of south-west New Zealand. And set to work on a series of poems using Arawata Bill as the personification, as the myth, of all of them.[8]

The result was the *Arawata Bill* sequence of 20 poems which were conceived, composed and published within a few months in 1953, just over five years after Bill O'Leary's death. As Denis Glover wrote to John Pascoe, "For better or worse this theme has taken hold of me. Bill is now up to twenty poems of greater or lesser length, and, I think, as good as anything I have ever done on a single theme."[9]

Some of the poems had nothing to do with Bill O'Leary. Poems such as *The Scene, The Search* and *The Crystallized Waves* were descriptions of the landscape and what Denis Glover called direct imagery of mountain, river and sea.[10] It was a landscape which owed as much to the Southern Alps of Canterbury so familiar to Glover from his climbing days, as it did to the country of North-West Otago and South Westland so familiar to Bill O'Leary.

Some were based on Bill O'Leary's life, for which the poet did his homework. John Pascoe had provided the raw material for two poems, *A Question*, and *Arawata Bill*, which was a re-arranged version of the original poem:

With his weapon a shovel
To test the river gravel
His heart was as big as his boots
As he headed over the tops
In blue dungarees and a sunset hat.

Wicked country, but there might be
Gold in it for all that.

Under the shoulder of a boulder
Or in the darkened gully,
Fit enough country for
A blanket and a billy
Where nothing stirred
Under the cold eye of the bird.

Some climbers bivvy
Heavy with rope and primus.
But not so
Arawata Bill and the old-timers.

Some people shave in the mountains.
But not so
Arawata Bill who let his whiskers grow.

I met a man from the mountains
Who told me that Bill
Left cairns across ravines
And through the scrub on the hill
– And they're there still.

And he found,
Together with a kea's feather,
A rusting shovel in the ground
By a derelict hovel.

It had been there long,
But the handle was good and strong.[11]

The background information Denis Glover had gleaned from his climbing mate was supplemented with a dossier of public library research material.[12] Much of this came from Alice McKenzie's *Pioneers Of Martins Bay*, which had been published six years earlier.[13]

Two of the poems were little more than poetic re-writes of passages from Alice McKenzie's book. The poem *By The Fire* was based on her account of the schoolboy bushrangers incident.[14] The poem *Incident* was based on her dubious account of the Cashmore story:

The constable said one day
'Wata Bill's too long away –
If you're going up by Deadmans Hut
Just look around for a bit.

The route he went,
You'll probably find him stiff
At the foot of a cliff,
Or dead in his tent.

But don't bury him,
Just cover him up.
Leave the body to me —
It's my duty to have the last look.'

When Cashmore saw two legs
Sticking out of a tent
With no camp smoke,
He dragged at them heavy-hearted.
And the sleeper awoke.[15]

Presumably for dramatic purposes, Denis Glover changed the location of the incident from the Pyke to Deadmans hut in the Hollyford.

Information from *Pioneers Of Martins Bay* found its way into several other poems, including *In The Township, Living Off The Land, To The Coast, Conversation Piece* and *The Little Sisters*.[16]

Background material also came from newspaper clippings, such as one in the *Free Lance* of 30 August 1939 which formed the basis for the poem *He Talks To A Friend*:

Hot springs I've found
Over on the other side,

And in the Red Hills
A lode of pure iron
Three feet wide.

Rubies and garnets
In odd patches,
And in one place shale oil
You can light with a box of matches.

The gold? Oh yes, I've got a good clue.
But you'd hardly expect
Me to tell you.[17]

Newspaper clippings also provided information for references such as the "seaboot full of gold" in the poem *To The Coast*.[18]

At the same time, Denis Glover ignored Alice McKenzie's statement that Bill O'Leary never smoked. In the poem *In The Township* he included the phrase "baccy in his pouch".[19] Pure invention too are various mentions of him getting drunk, cursing, even wearing "blue dungarees".[20] The same applies to references to "Paradise Pete" and "Lizzie the big blonde barmaid",[21] and all the words and thoughts and feelings Denis Glover put into Arawata Bill's mouth and head and heart.

This is not surprising, given the poet's "notion of trying to turn fact into poetic legend". *Arawata Bill* was not intended to be a sequence of poems which celebrated the life of a man called William O'Leary. Denis Glover didn't even mention his real name, instead using Arawata Bill, Wata Bill and Bill. William O'Leary became an archetypal figure. As Denis Glover wrote to John Pascoe:

Libraries and interloan files have been great help; but I have chosen to treat O'Leary as the symbol of all the great unknown explorers, prospectors, even mountaineers, who have been looking for something intangible round the next bend.[22]

In that sense it made no difference whether Arawata Bill was a real life character, such as another of Denis Glover's poetic subjects, his Banks Peninsula mate Mick Stimpson. Arawata Bill the man might just as easily have been another imaginary figure such as Harry of Denis Glover's *Sings Harry* fame. *Arawata Bill* the sequence of poems might just as easily have been about the real-life explorer Charlie Douglas or gold prospector Alphonse Barrington, who both rated a mention, or some fictitious Waiatoto Willie.

In the choice of Bill O'Leary, a crucial factor was his ready-made nickname. As Denis Glover wrote, he was taken with the name. The nickname in turn was crucial in Bill O'Leary's progress towards the status of national legend. As a *New Zealand Listener* magazine editorial observed, following the publication of the sequence of poems:

...perhaps it is more than chance which seems to be making William O'Leary the prototype. If there were few other good reasons his nickname would make him irresistible. The legend had already started when somebody first called him Arawata Bill.[23]

Denis Glover later wrote that, without thinking about it, he seemed to have contributed something to New Zealand folklore.[24] John Pascoe went further. According to him, the effect of his friend's sequence of poems was to give O'Leary "an assured place in the Valhalla of New Zealand explorers".[25] Furthermore, that largely through the sequence, Arawata Bill became for the intellectual New Zealand the symbol of inarticulate wandering, and of "man alone", as had the hero Johnson in John Mulgan's novel of that name.[26]

If so, it was not because Bill O'Leary fitted the Johnson mould of a misfit who was running away from something. That real life vacancy would be more aptly filled by Robert Monteguoy Nelson, alias Joe Driscoll. He became a legend when he deserted during the Second World War and spent five years hiding out from the military police in South Westland. After serving a year's hard labour he returned to South Westland where coincidentally he worked as a roadman on the Haast-Paringa track, the same job Bill O'Leary had done a couple of decades earlier.[27] Unlike the unreal Johnson and the real Joe Driscoll, Bill O'Leary chose to be a man alone in an environment in which he was at home. If he was running anywhere, it was towards something.

The *Arawata Bill* sequence of poems is regarded as one of Denis Glover's major achievements. It became a poetic best-seller. The first edition sold out in two weeks. A re-issue appeared in 1957. A high quality limited edition of 100 copies was published in 1981. It included half-a-dozen etchings by Gary Tricker illustrating some of the scenes described in the poems.[28] The sequence of poems also appeared in the 1964 and 1971 editions of *Enter Without Knocking, Selected Poems By Denis Glover*.[29] The poems were studied and dissected in schools and universities and in books and articles.

The sequence was not restricted to the printed page. The first Arawata Bill poem was broadcast on the New Zealand Broadcasting Service's programme *First Hearing*, along with a story about how it was written.[30] Shortly after Denis Glover wrote and published *Arawata Bill, A Sequence Of Poems* in 1953, it was broadcast as a reading for four voices on the New Zealand Broadcasting Service's non-commercial radio stations. To promote the programme, 'Arawata Bill' and Dolly appeared on the front cover of the *Listener*, courtesy of Thelma Kent, and Arawata Bill was the subject of an editorial and article.[31]

Nearly two decades later, the sequence of poems, along with several others, was recorded by Denis Glover and released as a long-playing record entitled *Arawata Bill And Other Verse*. It came complete with a cover featuring Arawata Bill and Dolly, again courtesy of Thelma Kent. The record also contained two of the poems, *Camp Site* and *Soliloquies*, set to music and sung by Les Cleveland, with guitar backing.[32]

Les Cleveland had previously set to music a selection of the poems for his wife Mary to use as material for the teaching of New Zealand poetry in various secondary schools. These were later complemented by a tape recording of Denis Glover reading the poems. Out of these initial experiments came the long-playing record.[33] Later still, Les Cleveland included one of the poems, *Arawata Bill*, in his *The Great New Zealand Songbook*.[34]

The poems were given the classical treatment too. In 1971, for example, a Canterbury University music student, Ross Harris, arranged *Arawata Bill* for baritone and orchestra. Denis Glover was not impressed with what had been done to his words, walking out during a performance of the work to which he had been invited in Christchurch.[35] Similarly, in 1983, after the poet's death, West Coast-born Wellington composer and music teacher Dorothy Buchanan wrote a composition for choir and piano based on three of the *Arawata Bill* poems, *The Scene*, *The Search* and *The Prayer*. Eight years later she wrote a solo piano version of *The Scene*, entitled *From The Mountains To The River*, which had been commissioned by New Zealand Symphony Orchestra concertmaster Isidor Saslav and his wife Ann.[36]

John Pascoe had been instrumental in Denis Glover's writing his *Arawata Bill* sequence of poems. On his return to Wellington, John Pascoe had attempted something himself. According to John Pascoe, his friend was "a bit indignant when he found that I too had written my free verse on the same subject".[37] It was also called *Arawata Bill*:

William O'Leary made his trails and cooled gold fever
in rivers and gorge.
He scratched the outcrops of basic rock intrusions,
dug alluvial gravel, sifted it and found a pass
over land to the sea.

The Tasman Sea flowed full to ragged cliffs and its tradition
gave granite hardness to resolution shrunk by hardship.

He smelt the sea and headed far up valleys
for solitude. Rare metal his aim on the rocks
over the land to the sea.

Before Bill came to traverse the Dart, the Pyke, the Cascade,
the Joe, the Barrier, the Jerry, the Woodhen, the silt of
the Williamson, the lush flats, the barren bluffs and
turning course of torrents, great peaks primeval raised up
in lofty quietness
over the land to the sea.

Before his boots had scraped the moss from the boulders,
explorer Douglas and the prospector Barrington
had fused their sweat in the sweetness of mist. The terror of time
and the anger of stonefall had marked the passage of all men
over the land to the sea.

Yet the terror of time had been chafed by the patience of reason,
the anger of stonefall had melted before the agility
of men who found no bush hostile. Camp oven, crowbar,
flour, bird-flesh, sleep made travel possible
over the land to the sea.

What did Bill think of Christ? Was he Saviour or used in epithet hot
to exorcise the scrub of animate devilment?
Bill died in Catholic sheets. Were memories strong of flight
from storms, fights with floods and snow that fell
over the land to the sea?

The shovels have rusted, the improvised huts have rotted,
men go the way of all men and perish their tea leaves
on mica schist flung. Their names a legend in the hills of no legend,

their footmarks buried by leaf mould and the mating of deer trails over the land to the sea.[38]

The poem was broadcast along with Denis Glover's initial *Arawata Bill* poem in *First Hearing* in 1953 and then published in the *New Zealand Poetry Year Book* of that year. It was later reproduced by John Pascoe in his book *Great Days In New Zealand Exploration*, published in 1959.[39]

Denis Glover and John Pascoe were not the only ones to follow in the poetic tradition of Dinny Nolan. John Pascoe's experience of finding the cairns had moved him and fired Denis Glover to write poetic tributes. Another climber who followed in Bill O'Leary's gumboot steps was similarly moved.

In 1982 Mount Aspiring National Park ranger Brian Ahern completed a return crossing of O'Leary Pass from the Dart to the Joe with fellow climber Geoff Spearpoint. According to Brian Ahern, it was a trip both had secretly wanted to do for many years. For them the seldom-used pass carried a special mystery, built up of tales from people who had rubbed noses with the steep western flank and had experienced their share of excitement on the bluffs and precipitous ledges. Such tales had grown with each telling until many regarded the crossing with fear and trepidation.[40]

At the end of the trip, Brian Ahern penned his own Arawata Bill tribute, which was published in the *Mount Aspiring National Park Newsletter* that year:

The Stones still Rattle in the creek
As the water from the Nor-westers speak
Pick and Pan shovels rusted on river terrace
now moss covered
The make shift shelter from the storm now gone is
but a dark stain in the greyness of the sand
Many feet now have trod your ways in a
Mountainous hilly land etched deep
by scar of legend, lives on

O'Leary's fame not from spoken word
but from a mystery alone.[41]

When the Southland Tramping Club proposed erecting a memorial head-stone over Bill O'Leary's grave in 1958, an *Otago Daily Times* columnist offered support. Referring to the Glover and Pascoe poems, the columnist added:

> When this old wanderer died a few years ago he was already a legend in the land. In an earlier generation he would no doubt have become the subject of a rough-shaped ballad or two – a suitable tribute which he might have appreciated. But he has had poetic justice passed to him – well, and at length – by two of our leading poets. So he has his memorial in literature, a recognition which has been given to few in this country.[42]

A more sober comment appeared in the *Truth* article which was written as part of the publicity surrounding the 1958 fund-raising appeal:

> Significantly, he is becoming something of a New Zealand myth. His name crops up these days in sophisticated North Island circles as a symbol of the redblooded past and man-against-the-frontier tradition, with accompanying poetic rhapsodies which would have made the old joker cackle to himself in the bush for months if he could have heard such things.[43]

CHAPTER SIXTEEN

PADDOCKS
& PAINTINGS

$\mathscr{T}$ributes and legend building were not restricted to the poetic variety following Bill O'Leary's death. The publicity surrounding the 1958 fund-raising appeal had boosted the Arawata Bill legend in New Zealand. It was given wider impetus in 1966 when Rupert Sharpe's account of his 1949 mustering trip with Davie Gunn was posthumously published in London. In *Fiordland Muster*, Rupert Sharpe reproduced the story Davie Gunn told him of Bill O'Leary's route over the Humboldt Range, during which he carried only a billy full of porridge, and spent a night in an ice cave on the tops.[1]

Rupert Sharpe then exclaimed to his international readership:

Well, Holy Moses! I've met a few New Zealanders that I would describe as middling tough, but imagine days of travel, over the most desperate country you have ever seen, on a handful of porridge in the morning and another handful at night. I shouldn't think you would bother about a midday handful. No tea, no tobacco, no blankets – and you spend one night in an ice cave which wouldn't be much below 5000 feet. It was probably someone like Arawhata Bill who started the Himalayan legend about the Yeti. I'll bet he was too wild to let any chance-met musterers or alpinists get a good look at him, but they might find tracks or a strong scent.[2]

The legend gained further momentum during 1972. The first burst of publicity came in response to a request from Murray Gunn for information about Bill O'Leary for display in Murray Gunn's Hollyford Museum. The appeal was made through Jim Henderson's popular 'Home Country' column in the *New Zealand Farmer* magazine. As a result, the column ran stories about Bill O'Leary for several issues, including contributions from Denis Glover and John Pascoe, Tom Carter, Dan Greaney and Rosie Grant.[3]

During that year, Bill O'Leary also became the subject of a radio documentary which the re-named New Zealand Broadcasting Corporation played on its National Programme. The *Spectrum* documentary was described as a re-creation of the life of the near-legendary prospector and solitary through the eyes of those who knew him, and through the poetry of Denis Glover. The half-hour programme, which was written and produced by Alwyn Owen, was based on interviews recorded during the 1960s and 1970s.[4]

The following year an article devoted to Bill O'Leary appeared in the seven-volume *New Zealand's Heritage, The Making Of A Nation* series. The article was written by Alwyn Owen and based on the *Spectrum* documentary material. Bill O'Leary didn't make the *Dictionary Of New Zealand Biography* published during his lifetime to mark the nation's centenary in 1940. Fifty years later he was regarded as being significant enough to appear in the *Dictionary Of New Zealand Biography* published to mark the nation's 150th anniversary. The essay was again written by Alwyn Owen.[5] A television documentary was also considered, but nothing came of it.

The Arawata Bill legend was continued in other ways, however. At Murray Gunn's museum, photographs, newspaper clippings and book extracts continued to be displayed alongside the pick head salvaged from the fire which destroyed the other personal mementoes in 1990.

The legend was also perpetuated in Department of Conservation visitors' centres on the fringes of Bill O'Leary's territory. The centre at Haast included a prominent display featuring a sketch copied from one of the Thelma Kent photographs. It was accompanied by an obligatory extract from Denis Glover's sequence of poems and a caption which proclaimed Bill O'Leary as "The most renowned European resident of the great South West". On the other side of the Main Divide, the visitors' centre at Glenorchy also included a tribute to Bill O'Leary. It comprised a Thelma Kent photograph, Denis Glover extract, and caption which described Bill O'Leary as "one of the most doggedly optimistic of the solitary gold-miners prospecting between here and the West Coast." The Department of Conservation centre at Wanaka paid tribute to his crossing of O'Leary Pass.

It was a similar story at Arrowtown's Lakes District Museum, with its display of Thelma Kent photo, Denis Glover extracts and tribute. It informed visitors that the poems were based on incidents in the prospector's life, "but in a wider sense they are meant to personify in Arawata Bill, all the unknown prospectors who essayed rough and wicked country that is not yet fully explored."

In nearby Queenstown there were tributes too. One of the streets in a 1974 subdivision was called Arawata Terrace. The Sunshine Bay subdivision was developed by Bill and Elsie Thompson's son Bob out of part of their farm where Bill O'Leary had spent a couple of years. Bob Thompson made sure the street was named in honour of his parents' guest.[6]

The street names didn't stop there. Off Arawata Terrace, a cul-de-sac called O'Leary's Paddock was named after him by the Queenstown Borough Council. It was shades of Charlie Douglas and Learys Pass. The Council had a policy of naming new streets after early prospectors, pioneers and surveyors. It decided on O'Leary's Paddock in the mistaken belief that Bill O'Leary used to paddock his horse there. Trouble is he never had a horse during his days in Queenstown and lived some distance away.[7]

More transient tributes to the town's former resident included a hotel bar named after him. The former Travelodge Hotel in Queenstown included the O'Leary Room, which featured among its historical paintings one the hotel commissioned Lloyd Veint to paint. The former Glenorchy scheelite miner knew Bill O'Leary when they stayed together at the Mount Earnslaw Hotel, before Lloyd Veint went on to buy Paradise House and then run Arcadia Station at the head of Lake Wakatipu. The large oil painting was based on the Thelma Kent photograph of prospector and pack-horse, with the addition of shovel and gold pan to Dolly's load.[8]

When the Travelodge became the Park Royal, the O'Leary Room went by the boards. The painting and a written tribute ended up in pride of place in the Rosewood Country Club Restaurant opened by Eddie and Sylvia Woodhouse at Middlemarch, inland from Dunedin.

Legends tend to gain their own momentum. Dozens of books have mentioned Bill O'Leary or Arawata Bill, usually in passing and often in conjunction with Thelma Kent's photographs and Denis Glover's poems. One annual publication appeared to take literally the saying that he had been immortalised by Denis Glover. *Stone's Otago And Southland Directory* continued to included the entry "O'Leary Wm, prospector, Glenorchy" from 1942 through to its final edition in 1954, 12 years after he left Glenorchy and seven years after his death.[9]

Similar errors have found their way into other written material about Bill O'Leary, who must take some of the responsibility. Tom Bryant senior met him at Glenorchy one day. It was Bill O'Leary's birthday. He told Tom Bryant he was 75, the same age he had been the previous birthday.[10]

It's understandable then that the man who was born in 1865 and died at the age of 82 is variously said to have been born in 1864 and 1866, and to have died between the ages of 80 and 93. Even his death notices were a couple of years out, describing him as "aged 80 years".[11] Similar confusion has existed regarding his place of birth. Probably because of his surname, it's been assumed that he was born in Ireland.[12]

The exaggeration of Bill O'Leary's age, that for example he was 84 when he came down out of the mountains,[13] has tended to fuel the legend. The reverse has also occurred. Some who have written about him have done his name a disservice and detracted from the legend. The most serious and damning case has been to confuse him with an earlier West Coast character known as Maori Bill, whose real name was probably Timothy Kelly.

The latter was an immigrant from Northern Ireland who had apparently deserted from the so-called Maori Wars of the 1860s after being suspected of shooting an officer. He lived for some time under the name of Hill among the Maori inhabitants at Bruce Bay. That was until a visitor who had also fought in the Wars asked him if his name was Timothy Kelly. He denied it, but a short time later he made his way down the coast to Browns Refuge and then on to Big Bay about 1878. There he took possession of the house left vacant the previous year when Andy Williamson drowned with several others when his whaleboat was wrecked on a trip between Big Bay and Jackson Bay.[14]

At Big Bay he called himself William Thompson but was usually known as Maori Bill. For years he avoided all company. When a vessel called he disappeared into the bush. In the absence of boots he wore home-made sandals of plaited flax. He never bought stores but lived on fish and birds and towards the end on food supplied by the McKenzie brothers. They also arranged for him to receive the old age pension, for which he astonished them by signing his name Timothy Kelly.[15]

Andy Williamson's house had a good supply of books, including a complete set of the works of William Shakespeare. Anyone unexpectedly going near the house would hear the seemingly well-educated man marching up and down reciting Shakespeare's plays, most of which he knew by heart. He would also scare away stray prospectors by playing the madman and with talk of shooting and of burning down their huts. This gave him both solitude and whatever the frightened prospectors left behind.[16]

After about 40 years at Big Bay, and thought to be over 90 years of age, the feeble Kelly was taken by steamer to the Old Men's Home in Kumara, a move arranged by the McKenzie brothers. He died there in 1918 and was buried under the name of William Edgar Thompson.[17]

Bill O'Leary met Timothy Kelly on at least one occasion at Big Bay, during the 1912 expedition with Jack Aitken and James Webster, who described him as "Mr Kelly-Thompson". Arawata Bill and Maori Bill were definitely not the same man.[18]

Alice McKenzie in her 1947 *Pioneers of Martin Bay* wrote about both men, but made a clear distinction between them.[19] So did John Pascoe in his *Mr Explorer Douglas*, published a decade later, although he referred to them in consecutive sentences:

Maori Bill was an Irish deserter from the Maori Wars, who had a hide-out in the Cascade Valley, and clothed himself in kiwi skins, only coming out for stores and tobacco. "The immortal wild man Maorie Bill...actually planted a patch of strawberries, the only piece of work he was ever known to do." Douglas never met William O'Leary, Arawata Bill as he was called, though he referred to his transalpine crossing from the Dart.[20]

The quote about the wild man and the strawberries came from Charlie Douglas.

In 1976, in his *Jacksons Bay, A Centennial History*, Irvine Roxburgh took John Pascoe's account of the two men and turned them into one:

O'Leary usually went by the name of Arawata Bill when he appeared in town – but this he seldom did. As Irish as his name, O'Leary had deserted the British forces during the Maori Land Wars and had chosen the Cascade Valley as his hideout, clothing himself in skins and venturing into civilization only when in need of stores or tobacco. It is said that his planting of a bed of strawberries was the only work, apart from roaming and fossicking among the mountains, that he had ever been known to do – but such a life was surely hard work in itself.[21]

The unfortunate mix-up took on a life of its own. Three years later, for example, Neville Peat in his *Cascade On The Run* compounded the error:

William O'Leary, an Irish-through-and-through roadman, ferryman and prospector, was also known as Maori Bill because he deserted from the British forces in the Maori Wars and was sometimes seen clothed in kiwi skins and feathers. ...He had a hideout in the Cascade Valley where he is credited with putting in a patch of strawberries. Detractors were quick to point out the strawberries represented his only work beyond the fabled Arawata Bill vocation – fossicking. O'Leary rarely came to town.[22]

Associated with this confusion is the assertion made from time to time that Bill O'Leary adopted his hermit existence because he was a deserter not during the Land Wars, but during the First World War. It was said that he was one of the shirkers who hid out in the area between Lake Wakatipu and the West Coast after conscription was introduced in 1916. It's an assertion without foundation. The 50-year-old Bill O'Leary had no reason to hide. He was too young for one war and too old for the other.

CHAPTER SEVENTEEN

EVERLASTING WHISPER

*T*here's no alchemist's formula for making universal or New Zealand folk heroes and legends. If there were, the indigenous variety would probably include many of the elements Bill O'Leary possessed.

An independence and simplicity which enabled him to live alone for long periods in remote places. A strength and determination which enabled him to spend weeks moving heavy loads over long distances and up difficult gorges and over dangerous passes. An unaffected eccentricity which saw him tackling snow-covered mountain slopes while dressed in three-piece suit and thigh gumboots, and wielding a miner's pick. A genuine and endearing modesty which prevented him from seeing anything remarkable in his endeavours. He was in his own eyes that very much over-estimated man.

In the absence of scandal or a contempt for authority which served other legendary characters so well, a basic honesty and decency and personal morality. In the absence of a short life or dramatic death, a longevity which enabled him to continue his independent, determined, eccentric, self-effacing way of life into old age.

In Bill O'Leary's case, the long life was particularly important. At the time of his retirement as West Coast ferryman in his early sixties, he had achieved something of a name for himself in South Westland. Most of the exploits and human contacts which elevated him from the status of local identity to national folk hero and legend occurred during the final two decades of his life. Had he drowned in the Waiatoto River on his last ferry crossing, he would probably have remained that local identity.

Other elements might include a history that is oral and open and ripe for

Bill O'Leary looks to the beckoning hills in Queenstown, 1940.

embellishment. Add a touch of the romance of gold and buried treasure and a lost ruby mine and the intrigue of a murder mystery. Add a pinch of metaphor in a lifelong search for a pot of gold at the end of a rainbow. A nickname such as Arawata Bill helps too.

The right ingredients, however, are no guarantee of success in making a legend out of a rabbiter, bushman, station hand, trackman, ferryman, prospector, treasure-seeker, rouseabout and pub cowboy. Others shared his traits but never achieved his status beyond their immediate circle or their own generation. The essential ingredient has to be luck. Something as simple as leaving a pile of stones and a miner's shovel at the right place at the right time. Then having the fortune denied him as a prospector to have a well-known poet write a major collection of poems originating in a chance remark about the accidental rediscovery of the stones and the shovel.

Bill O'Leary became the legendary Arawata Bill despite himself. It's not something he sought. For most of his 82 years he chose to quietly roam the rugged back country of North-West Otago and South Westland. He didn't make a fuss. He simply got on with it. Unlike many of the explorers, surveyors and prospectors who had come before him, he never kept a record of his wanderings. The only known records to survive his death were his letters to Charlie Haines and Bob Moodie. Other signs of his existence were limited to a blaze on a tree here, a stone cairn there, the remains of a rusty shovel or kettle or of a camp site or terraced cave floor, a pick head in a museum, a headstone in a graveyard.

It's hardly surprising. In a purely material sense Bill O'Leary had few possessions, didn't need them, didn't want them. When Thelma Kent photographed him on the river rocks in the late 1930s, she photographed nearly all he ever owned. The clothes in which he stood, his horse and the gear and tucker packed in sacks on the horse's back. Somewhere he would have cached his pick and shovel and panning dish and maybe some emergency supplies. He never owned a piece of land or a house or anything he couldn't pack on a horse or carry on his back.

There are many explanations offered for his choice of this way of life. The over-simple answer is chronic gold fever. He was convinced there was considerable gold to be found and if he kept on fossicking long enough he would come across it.[1]

If he was solely interested in making his fortune, there were more likely places he could have gone. Gold was an excuse. As his *Evening Post* obituary mused:

...always behind his journeying there seemed more than just the mere lust for riches, something greater; that 'everlasting whisper' which calls men again and again to the heart of the beckoning hills.[2]

If there was a single attraction it was the hills themselves and the way of life they afforded, not the gold they might have contained. As Harry Fortune wrote in the *Wide World Magazine* after Bill O'Leary's death:

It was not the hope of riches that kept Bill in the mountains as old age crept upon him; it was nothing more or less than love of an intensely healthy existence, in which all days were the same and yet never the same, and in which he was free to come and go as he pleased, calling no man master.[3]

The simplicity of Bill O'Leary's answer to Dan Greaney in Queenstown in 1940 perhaps best sums it up:

We carried on with the conversation and I said to him, "Bill, a lot of people reckon that the motive for you living in that wild region so long was just your love of gold, and probably those rubies, and the imaginary Cascade treasure." He laughed. He said, "Oh yes, a little, but," he said, "I really loved the bush and the mountains. That was the main reason."[4]

APPENDIX

Kinlock.

Sunday September 10th 1937

    Dear Robert Moody
Just a few lines to hear how
you are doing. old T Bryant
went away with you or after you
and he has not come Back yet
so iam wondering how you are
getting on weather is Bag here just
now, harry is up the Dart Packing,
guide seater. and two Morelles. from Manopori
lake with him with their Riffles tellicopes sights.
on them, so the Dears will get a rough time
I am feeling the effect of being to close to the
red Beech, heavy snow on the Mountanes now
I wont be able to get away before november,
hoping you have not lost the sight of your Eye.
hoping to hear from you soon, Sheelight is going up
a lot of work going on onthe west coast at, Bruce Bay.
 just now. To Glenorchy boyes working there
(Co. Mr T. Bryant)
    Yours Truly Wm J. O'Leary
    Kinlock Lake Waikatipu

Kinlock October 10th 1927 [1937]
    Dear Robert
I Received your welcom letter
glad to here you are well as leves
me at present, bad colds going about,
am still at the garden going.
fencing Monday, to much snow
on the tops yett, will take a
load of tucker up to the hut
in Dart
at the end of the month.
will be living next month
for good, still gathering things
for trip, they, are trying to stop me
they made a raid on the Dear
the skins went ten shillings each
from the Dart harry Bryant got
16 seater 8. and Morells about
20 skins
    Yours Truly
    W J O'Leary
    Kinlock Wakatipu

October 1st 1939
Mr Robert Moody
Dear Sir
    Glad to here
from you things is verry quit
here I have been twelve mounts here
now, to clos to queenstown to save much
i have been telling some that i was going

to walk to the Exhibition up the coast, but
it wont be this year, this war is bad for us
Miss Bryant & Harry is running the Bording
hous, there is no one travling and the Bluff boat stoped
littelton Boat stoped subremean smewhere.
things looking bad for Exhibition,
When you get down to Burk River
You will be in good reafing and gold bearing
country heavy gold in Burk River
You Might Sea Me Passing through at christmas time
 Yours Truly
 W.J. O'Leary queenstown
 Arawata Bill

queenstown
Wednesday. Nov 15th 1939
Mr. Robert Moodey
Dear Sir
 am posponing trip
till christmas time or after
a trip from Invercargill for a week in
Wellington is advertised. costing
Eight Bound ten shillings
its a cheap trip, in southland paper.
otago daily times wednesday nov 15 has a old
photo of me three or four years old
things Verry quit here. verry verry few coming
to queenstown this christmas. Exhibition doing
well. dont know what I. will be doing till i
come back from Wellington
hopping Youare all Well. am 12. 1/2 stone now

Yours Truly
Wm J. O'Leary
queenstown
P.O.

Wenesday March 20, 1940
Dear Robert
I received your welcom letter
two dayes ago, had fine time at
Exhibition. well worth seeing. verry good
it onely cost me eighteen Pound
and am glad i took the trip up
it cost me four Pound to go up (2 class
and the same to come down. the Boat was
verry crouded Both up and down trip
have shifted up to thomsons homested
and have a fine little cotage with stove in it
we are hoping for a fine day for rowing
on saturday.
    Yours Truly. Arawata Bill
    Wm J. O'Leary queenstown

I will stop here till october now
to colect a fue bob again if i can
or shift up to twelve mile and stop
there glade to see you any time you
are up this way, In verry good health
hoping you are the same, queenstown
is a healthy Place to live in
    Wm J. O'Leary.
    queenstown

NOTES

Chapter 1: Wetherstons And Wagging

1 Wetherstons is the official spelling, although the name has apparently never been considered by the New Zealand Geographic Board and there have been several different spellings over the years. New Zealand Geographic Board Records, New Zealand Geographic Board, Wellington.

2 W.R. Mayhew, *Tuapeka, The Land And Its People, A Social History Of The Borough Of Lawrence And Its Surrounding Districts*, Otago Centennial Historical Committee, Dunedin, 1949, pp 29-31.

3 *Otago Colonist*, Dunedin, 4 Oct 1861, p 5. Passenger Lists Victoria Australia Onwards To New Zealand Ports, 1852 Onwards, Part 3, 1861-1865, 1861, no 4068, Otago Settlers Museum, Dunedin. The passenger list names the ship as *Empress Of The Sea*.

4 Mayhew, *Tuapeka*, p 213.

5 Death Records, Hokitika, 1900, no 45, Births Deaths And Marriages Central Registry, Lower Hutt. "Marriage Notices, 1861-1869, Tokomairiro", no 55, Births Deaths And Marriages Records, District Court, Balclutha. S.M. Dalley To Author. Timothy O'Leary's death records give his place of birth as St John, New Brunswick, although according to family tradition he was born on nearby Prince Edward Island.

6 Dalley To Author. *List Of Immigrants, Debtors To The Provincial Government Of Otago For Passage Moneys, 1869*, p 12, nos 2806-8. *Otago Colonist*, 29 Dec 1862. *Otago Daily Times*, Dunedin, 27 Dec 1862, p 4. The 1869 debtors' list gives Bedilia's name as Benelia.

7 Contemporary terms such as miles have been retained for measurements and currency.

8 "Marriage Notices", "Register, Milton 1", no 6, Balclutha District Court. *Otago Daily Times*, 2 Sep 1863, p 4.

9 "Birth Register, Tuapeka District, 1865", no 234, Central Registry. *Otago Daily Times*, 29 Jun 1946, p 8. When his father had the birth registered nearly two months later, he told the registrar his son's name was William, but elsewhere his full name was usually given as William James O'Leary and it's the name by which he went.

10 "Baptism Book For The Lawrence Mission", Catholic Church Otago-Southland Archives, St Joseph's Cathedral Presbytery, Dunedin. "Birth Index", 1864-79, Registrar-General's Office. Dalley To Author. "New Zealand Public School Register Of Admission, Progress And Withdrawal, Lawrence District High School, From 23/1/71 To 10/2/90", Lawrence Area School Papers, 129/89, Hocken Library, Dunedin. *Tuapeka Times*, Lawrence, 8 May 1873.

11 Catholic Church, *The Lode Of Faith, The Church In The Tuapeka, St Patricks Lawrence, Centennial Souvenir, A Diocesan Centennial Publication*, Lawrence, 1972, pp 16-17. Mayhew, *Tuapeka*, pp 176-7, 213.

12   Catholic Church, *Lode*, pp 16-17.

13   "General Applications For Gabriels Gully, 1862-1863", "General Applications For Gabriels Gully, 1863-1864", "Register Of Mining Applications, Record Book, 1874-79", Warden's Court, Lawrence, D98 Arc G7, National Archives Of New Zealand, Regional Office, Dunedin.

14   Mayhew, *Tuapeka*, p 218.

15   Dalley To Author. J. Cron, in A. Owen, "Scripts For Arawata Bill Broadcast On NZBC", 74-123, Alexander Turnbull Library, Wellington. Murray Gunn also has copies of most of the Scripts at his Hollyford Museum. The Scripts have been corrected against the few edited excerpts from the original recordings to have survived.

16   A.A. Cron To Author.

17   A. McKenzie, *Pioneers Of Martins Bay*, Southland Historical Committee, Invercargill, 1947, pp 77-8.

18   *House Of Representatives, Province Of Otago, Roll Of Electors For Bruce Electoral District For The Year Ending 30th September 1866*, p 6, no 379. *Tuapeka Times*, 8 May 1873.

19   Mayhew, *Tuapeka*, p 61.

20   Cron, "Scripts".

21   Dalley To Author.

22   Dalley To Author. *House Of Representatives, Electoral Roll, List Of Persons Qualified To Vote At The Election Of Members For The House Of Representatives For The Westland District, 1890, Hokitika Electoral Roll*, p 25, no 1442. *1900, Electoral District Of Westland, General Roll Of Persons Entitled To Vote For Members Of The House Of Representatives Of New Zealand*, p 56, no 3094.

23   Dalley To Author.

24   Dalley To Author. "School Register", Lawrence. *Stone's Otago And Southland Directory*, Dunedin, 1888, p 369, 1889, p 384, 1893, p 406, 1897, p 459.

Chapter 2: Passes And Prospecting

1   Dalley To Author.

2   Cron, "Scripts".

3   *Otago Daily Times*, 29 Jun 1946, p 8. See also C.H. Fortune, "The Passing Of 'Arawata Bill'", *Wide World Magazine*, London, vol 101, no 605, Oct 1948, p 310. *Evening Star*, Dunedin, 10 Nov 1947, p 6. *Otago Daily Times*, 11 Nov 1947, p 6.

4   *Otago Daily Times*, 11 Nov 1947, p 6.

5   *Evening Star*, 10 Nov 1947, p 6. Fortune, "Passing", p 310. *Otago Daily Times*, 29 Jun 1946, p 8.

6   *1902, Electoral District Of Wallace, General Roll Of Persons Entitled To Vote For Members Of The House Of Representatives Of New Zealand*, p 60, no 3324.

7   *1908, Electoral District Of Wakatipu, General Roll Of Persons Entitled To Vote For Members Of The House Of Representatives Of New Zealand*, p 68, no 3349. P.M. Chandler, "Arawata Bill", P.M. Chandler Papers, MS 1270, 2/2/3, Hocken Library, Dunedin. T. Kennett, "Scripts".

8   D. Matheson To Chandler, 14 Jul 1958, Chandler Papers, 4/6/10.

9   Matheson To Chandler, 14 Jul 1958.

10 C.H. Fortune, "Arawata Bill", *Wide World Magazine,* vol 75, no 448, Jul 1935, p 289. *1896, Electoral District Of Waitaki, General Roll Of Persons Entitled To Vote For Members Of The House Of Representatives Of New Zealand*, p 50, no 2795. *Otago Daily Times*, 4 Jul 1947, p 6.

11   Arawhata is the original spelling, from the Maori ara, meaning path, and whata, meaning food store. Since it was gazetted by the New Zealand Geographic Board in 1990 it has also been the official spelling for the place, river, state forest, rock bivouac, glacier and saddle. Before then it was officially spelt by the commonly known name of Arawata. Geographic Board Records.

12   *Lake Wakatip Mail*, Queenstown, 22 Oct 1897, p 2.

13   "First Ascents And Explorations", *New Zealand Alpine Journal*, Dunedin, vol 6, no 22, Jun 1935, p 207. J.T. Holloway To Chief Surveyor, Dunedin, Jul 1937, no 63/38, Geographic Board Records.

14   "William Wilson Collection, Charles Douglas Exploration Material", C.E. Douglas Papers, MS 90, 44, Alexander Turnbull Library, Wellington.

15   "Exploration Material", Douglas Papers.

16   Holloway To Chief Surveyor.

17   I. Whitehead, "Ionia And The Arawata", *Alpine Journal*, vol 12, no 35, Jun 1948, p 207.

18   Chandler, "Arawata Bill".

19   B.N. Alexander, "An Historical Approach To The Olivines", *Alpine Journal*, vol 10, no 51, 1964, p 348. Fiordland National Park Map NZMS 273, Department Of Lands And Survey, Wellington, 1977, 44 30 00, 168 26 00.

20   Fiordland Map, Lands And Survey, 44 30 00, 168 01 00. G. Speden To Author. J. Hall-Jones, *Fiordland Placenames*, Fiordland National Park Board, Invercargill, 1979, p 61.

21   A. James, "Direct Crossings, Lake McKerrow To Kaipo Slip", *Alpine Journal*, vol 21, no 1, 1965, p 132.

22   C.L. Veint To Author.

23   Geographic Board Records. M.A. Gunn, "Arawata Bill", In J.H. Henderson (ed), *Open Country Muster, People And Places Out Of Town*, A.H. And A.W. Reed, Wellington, 1974, p 183. T.B. Richardson, "Legendary Man Of The Hills", *Southland Daily News*, Invercargill, 15 Nov 1947, p 10.

24   McKenzie, *Pioneers*, p 77. Jackson Bay is the official name, although it was commonly known as Jacksons Bay.

25   D. McKenzie, "1903 Diary", I.G. Speden, Lower Hutt.

26   Pyke is the official spelling, although it was commonly spelt Pike.

27   Waiatoto is the official spelling, although it was commonly spelt and pronounced Waitoto.

Chapter 3: Pigeons And Porridge

 1   Fortune, "Passing", pp 310-11.

 2   F. Heveldt, "Scripts".

 3   Richardson, "Legendary".

 4   C. McEwan, J. Swale, Kennett, "Scripts". *Otago Daily Times*, 29 Jun 1946, p 8. Richardson, "Legendary".

 5   Heveldt, "Scripts".

 6   McEwan, Swale, "Scripts".

 7   Speden To Author.

 8   Gunn To Author. Speden To Author.

 9   McEwan, "Scripts".

10   *Evening Star*, 10 Nov 1947, p 6.

11 McEwan, "Scripts".
12 *Otago Daily Times*, 29 Jun 1946, p 8.
13 A. Forbes, B. Heffernan, Cron, D. Greaney, E. Martin, F. Heffernan, G. Koch, J. Bourke, McEwan, R. Elliot, Swale, W. Clingin, "Scripts". Dalley to Author. D.J. Nolan To Author. Fortune, "Arawata", p 289. F.V. Yarker To Author. K. Nolan To Author. M.F. Shaw To Author. P.G. Sharpe To Author. R.F. Grant To Author. R.H. Bryant To Author. Richardson, "Legendary". Veint To Author.
14 Clingin, Cron, F. Heffernan, "Scripts". McKenzie, *Pioneers*, p 77. *New Zealand Free Lance*, Wellington, 5 Aug 1936, p 53. *Otago Daily Times*, 29 Jun 1946, p 8. *Otago Witness*, 28 May 1929, p 40. Richardson, "Legendary". Speden To Author. Veint To Author.
15 McEwan, "Scripts".
16 F. Heffernan, "Scripts".
17 E.V. Bryant To Author. F. Heffernan, Martin, "Scripts". K. Nolan To Author. Richardson, "Legendary".
18 Kennett, "Scripts".
19 F. Yarker To Author. J.L. McDonald To Author. McKenzie, *Pioneers*, p 77. *Otago Witness*, 28 May 1929, p 40.
20 *Evening Post*, Wellington, 25 Nov 1947, p 10. Greaney, "Scripts". N.I. Groves To Author. W.B. Beattie, *Bill Beattie's New Zealand*, Hodder And Stoughton, Auckland, 1970, p 38.
21 McDonald To Author. McKenzie, *Pioneers*, p 77. Clingin, McEwan, "Scripts". Richardson, "Legendary". Sharpe To Author.
22 Holloway, "Change And Decay In The Mountain World", *Massif*, Palmerston North, vol 9, Aug 1976, p 11.
23 Richardson, "Legendary".
24 R.J.D. Sharpe, *Fiordland Muster*, Hodder And Stoughton, London, 1966, p 63.
25 *Lake Wakatip Mail*, 22 Oct 1897, p 2.
26 Holloway, "Change", p 11.
27 Dalley To Author.
28 McDonald To Author.
29 *Free Lance*, 24 Jul 1935, p 20.
30 Richardson, "Legendary".
31 McEwan, "Scripts".
32 Heveldt, "Scripts".
33 B. Chapman To Gunn, M.A. Gunn, Papers, Hollyford Museum, Hollyford.
34 Greaney, McEwan, "Scripts". Groves To Author. Gunn To Author. Richardson, "Legendary".
35 T.A. Clark To Chandler, 19 Jan 1960, Chandler Papers, 4/6/10.
36 Speden To Author.
37 Greaney, "Scripts".
38 McDonald To Author.
39 Fortune, "Arawata", p 290.
40 M.N.D. Martin, In J.H. Henderson (ed), "Home Country", *New Zealand Farmer*, Auckland, vol 93, no 4, 24 Feb 1972, p 74.
41 Holloway, "Change", p 11.
42 Bourke, "Scripts".

43 Heveldt, "Scripts".

44 D. Galloway, "John Thorpe Holloway, A Memoir", *Alpine Journal*, vol 30, 1977, p 138.

Chapter 4: Farms And Ferrying

1 *1914, Electoral District Of Westland, No 1 Supplementary Roll Of Persons Entitled To Vote For Members Of The House Of Representatives Of New Zealand*, p 9, no 6300.

2 Cron, "Scripts".

3 Cron, "Scripts".

4 Cron, "Scripts".

5 Cron, "Scripts".

6 McKenzie, *Pioneers*, pp 65-6.

7 McKenzie, *Pioneers*, pp 65–6.

8 James Deans Webster, "Diary", 1912, Hocken Library, Dunedin.

9 Cron To Author. Cron, "Scripts", W.D. Nolan Senior To Chandler, 28 May 1958, Chandler Papers, 4/6/10.

10 *Canterbury Mountaineer*, Christchurch, vol 4, no 17, 1948, p 240. D. McKenzie, *Road To Routeburn*, John McIndoe, Dunedin, 1973, p 106. *Otago Daily Times*, 11 Nov 1947, p 6.

11 W.D. Nolan, *The Nolan Family, Pioneers Of South Westland, 1871–1993*, Thornton Publishing, Christchurch, 1993, p 53.

12 A. Mackey, Greaney, "Scripts". E. Robb, *Hawea Hospitality*, Eileen Robb, Wanaka, 1986, p 8. H.J. Buchanan To Author. K. Nolan To Author. M.E. Eggeling To Author. W.D. Nolan To Author.

13 K. Nolan To Author.

14 Greaney, Heveldt, "Scripts". K. Nolan To Author. Nolan, *Family*, p 54.

15 E.T. Stoop To Chandler, 15 Sep 1958, Chandler Papers, 4/6/10.

16 Nolan, *Family*, p 53.

17 Bourke, "Scripts".

18 F.J. Turner, "Geological Expeditions To Haast-Cascade Areas, 1929, 1930", pp 2, 5, Alan Frederick Cooper, Dunedin.

19 Turner, "Geological", p 7.

20 Bourke, "Scripts". Turner, "Geological", p 7. Nolan, *Family*, p 53.

21 Cron To Author. Bourke, Cron, "Scripts".

22 D. Nolan To Author.

23 D. Nolan To Author. K. Nolan To Author. Mackey, "Scripts". Nolan, *Family*, p 53. W. Nolan To Author.

24 Mackey, "Scripts".

25 Heveldt, "Scripts". Stoop To Chandler, 19 Sep 1958, Chandler Papers, 4/6/10.

26 E. Bryant To Author. Cron, Martin, "Scripts".

27 Cron, Elliot, "Scripts". Sharpe To Author.

Chapter 5: The Thousand Acres

1 McKenzie, *Pioneers*, p 77. *Otago Witness*, 7 Aug 1928, p 45.

2 Fortune, "Passing", p 311. *Otago Daily Times*, 29 Jun 1946, p 8.

3 Whitehead, "Ionia", *Alpine Journal*, vol 12, no 35, Jun 1948, p 202.

4 Cron To Author. Greaney, "Scripts". K. Nolan To Author.

5 T. Carter To Henderson, Mar 1972, Gunn Papers. See also Carter, "To Wanaka From The Olivines", *Alpine Journal*, vol 16, no 43, Jun 1956, p 364.
6 G. Mueller, "Reconnaissance Survey Of The Head-Waters Of Arawata And Waiatoto Rivers, Westland", *Appendix To The Journals Of The House Of Representatives*, Wellington, vol 1, 1885, C1A, p 26.
7 Nolan To Chandler.
8 Greaney, "Scripts". Nolan, *Family*, p 26.
9 Nolan To Chandler.
10 Alexander, "Historical", p 347.
11 D. Nolan To Author. Nolan, *Family*, p 26.
12 Speden To Author.
13 Nolan, *Family*, p 26.
14 *Evening Post,* 17 May 1929, p 11. *Hokitika Guardian*, Hokitika, 22 May 1929, p 2. *Otago Witness*, 28 May 1929, p 40.
15 *Evening Post,* 17 May 1929, p 11. *Hokitika Guardian*, 22 May 1929, p 2. *Otago Witness*, 28 May 1929, p 40.
16 *Otago Witness*, 20 May 1929, p 48.
17 *Otago Witness*, 7 Aug 1928, p 45.
18 Cron, Greaney, Mackey, "Scripts". Cron To Author. D. Nolan To Author.
19 Cron, Greaney, Mackey, "Scripts". Cron To Author. D. Nolan To Author.
20 Greaney, "Scripts".
21 Greaney, "Scripts".
22 K. Nolan To Author.
23 Cron To Author.
24 W.D. Nolan Senior, "Arawata Bill, Bill O'Leary", In Nolan To Chandler. The poem is reproduced courtesy of Dinny Nolan's sons Bill, Des and Kevin Nolan.
25 Stoop To Chandler, 15 Sep 1958.

Chapter 6: The Frenchman's Gold
1 Fortune, "Passing", p 312. J.P. Morant, "Frenchman's Gold", *New Zealand Railways Magazine*, Wellington, vol 12, no 2, 1 May 1937, p 96. Nolan, *Family*, p 69. Bill Nolan's account was also published in J.P. Nolan (ed), *A Saga Of The South, A Short History Of A West Coast Family, The Nolans*, Nolan Reunion Committee, Christchurch, 1983, pp 76-9.
2 Fortune, "Passing", p 312. Morant, "Frenchman".
3 Nolan, *Family*, p 69.
4 Nolan, *Family*, p 69.
5 Nolan, *Family*, p 69.
6 Nolan, *Family*, p 69.
7 Nolan, *Family*, p 69.
8 Morant, "Frenchman".
9 Fortune, "Passing", p 312. Morant, "Frenchman". Nolan, *Family*, p 69.
10 Fortune, "Passing", p 312. Morant, "Frenchman".
11 Nolan, *Family*, p 69.
12 Fortune, "Arawata", pp 290-1.

13 Chandler, "Arawata". Sharpe, *Fiordland*, pp 62-3.

14 Chandler, "Arawata". Sharpe, *Fiordland*, pp 62-3.

15 *Free Lance*, 13 Jul 1938, p 70. McEwan, Swale, "Scripts".

16 *Free Lance*, 13 Jul 1938, p 70. McEwan, Swale, "Scripts". McKenzie, *Pioneers*, p 39.

17 *Free Lance*, 13 Jul 1938, p 70.

18 McEwan, Swale, "Scripts". Richardson, "Legendary".

19 McKenzie, *Pioneers*, p 39.

20 Fortune, "Passing", p 312.

21 Kennett, "Scripts". Sharpe, *Fiordland*, pp 62-3.

22 E. James, "Exploration In Unknown Otago, New Lakes And Routes Discovered, 1. The Journey To Martins Bay", *Otago Daily Times*, 3 Sep 1930, p 14.

23 Clingin, "Scripts".

24 Fortune, "Arawata", pp 290-1. Groves To Author. Gunn To Author. J.H. Beattie, *Far Famed Fiordland*, Otago Daily Times And Witness Newspapers Co, Dunedin, 1950, pp 84-5. Morant, "Frenchman". Richardson, "Legendary". Sharpe, *Fiordland*, pp 62-3. Shaw To Author. T.J. Thomson To Author.

25 Nolan, *Family*, p 67. The same account was also published in Nolan, *Saga*, pp 67-8.

26 Nolan, *Family*, p 67.

27 Nolan, *Family*, p 67.

Chapter 7: The Lost Ruby Mine

1 R.S., "A Secret Of The Olivine Range", *Otago Witness*, 4 Feb 1897, pp 52-3.

2 Richardson, "Legendary".

3 B.A. Jackson, "Otago's Lost Ruby Mine", *Otago Daily Times*, 12 Aug 1967, p 13. Chandler, "Arawata".

4 Chandler, "Arawata". Jackson, "Lost".

5 Richardson, "Legendary".

6 Chandler, "Arawata". Chandler To Jackson, 18 Mar 1966, Chandler Papers, 4/6/10. Jackson, "Lost".

7 Chandler, "Arawata".

8 W.J. O'Leary To C. Haines, 12 May 1930, In Chandler, "Arawata", Henderson, "Home", *New Zealand Farmer*, vol 93, no 6, 23 Mar 1972, p 69.

9 Gunn To Author. *Weekly News*, Auckland, 21 Oct 1959, p 7.

10 James, "Journey".

11 Jackson, "Lost". R.W. Duncan, "Arawhata Bill And The Ruby Mine", *Weekly News*, 8 Jul 1968, p 35.

12 Duncan, "Arawhata". Jackson, "Lost". *Southland Times*, Invercargill, 16 Mar 1968, p 16, 27 Dec 1971, p 5, 17 Jun 1982, p 22.

13 *Otago Daily Times*, 29 Jun 1946, p 8.

14 Grant To Author.

15 M. Huthnance To Chandler, 24 Jul 1958, Chandler Papers, 4/6/10.

16 Jackson To Chandler, 22 Apr 1968, Chandler Papers, 4/6/10.

17 Duncan, "Arawhata". M. Pickering, *The Southern Journey*, Mark Pickering, Christchurch, 1993, p 84.

18   Koch, "Scripts".
19   Gunn To Author. Jackson To Author. *Southland Times*, 16 Mar 1968, p 16.

Chapter 8: Miners And Mosies
 1   Beattie, *New Zealand*, p 38. James, "2, Over The Red Mountains", *Otago Daily Times*, 4 Sep 1930, p 7.
 2   Richardson, "Legendary".
 3   Beattie, *New Zealand*, p 38.
 4   Beattie, *New Zealand*, pp 38-40. James, "Mountains".
 5   Beattie, *New Zealand*, pp 39-40.
 6   Kennett, "Scripts".
 7   Martin, "Scripts".
 8   McEwan, "Scripts".
 9   Gunn To Author. McEwan, "Scripts".
10   Gunn To Author. McEwan, "Scripts".
11   McEwan, "Scripts".
12   L.F. McCurdy, "Invercargill Trip Book", Ann Irving, Waimatua. Southland Tramping Club, *Southland Tramping Club, 25th Jubilee, Southland Tramping Club Inc 1947-1972*, Southland Tramping Club, Invercargill, 1972, p 22.
13   Speden To Author.
14   Chapman To Gunn, Gunn Papers.
15   F. Yarker To Author.
16   McCurdy, "Trip". Southland Tramping Club, J*ubilee*, p 22.
17   B. Heffernan, "Scripts".
18   Beattie, *New Zealand*, p 39. Fortune, "Arawata", p 290.
19   A. Anderson To Author.
20   Richardson, "Legendary".
21   Richardson, "Legendary".
22   F. Heffernan, "Scripts".
23   B. Heffernan, Clingin, Cron, Elliot, F. Heffernan, Forbes, Kennett, McEwan, Swale, "Scripts". Dalley To Author. D. Nolan To Author. Grant To Author. Gunn To Author. McDonald To Author. R. Bryant To Author. Speden To Author. Veint To Author.
24   Richardson, "Legendary".

Chapter 9: Gumboots And Glaciers
 1   E.H. Sealy, "Exploration In New Zealand In 1937, Discovery Of The Olivine Ice Plateau", *Christchurch Star-Sun*, Christchurch, 15 May 1937, Magazine Section p 1. Holloway, "On The Olivine Ice Plateau, Great Playground Of The Future", *Evening Star*, 20 Feb 1937, p 26. Sealy, "In Unknown Country, The Olivine Ice Plateau, A 'Forgotten' River", *Otago Daily Times*, 13 Mar 1937, p 5.
 2   Sealy, "Exploration", "Unknown".
 3   Holloway, "Plateau". See also Holloway, "Olivine Bound, A Three Month's Record", *Alpine Journal*, vol 7, no 24, Jun 1937, p 60.
 4   Holloway, "Change", pp 10-11.

5 *Mataura Ensign*, Gore, 9 May 1973, p 2. McKenzie, *Road*, p 106. *Southland Times*, 27 Dec 1971, p 5, 17 Jun 1982, p 22.

6 J.M. Childerstone, "Museums, Hollyford Memories", *New Zealand Geographic*, Ohakune, no 1, Jan-Mar 1989, pp 112-13.

7 Galloway, "Memoir", p 138.

8 F. Heffernan, "Scripts".

9 Holloway, "Change", pp 10-11.

10 Galloway, "Memoir", p 138. Holloway, "Change", pp 10-11.

11 Galloway, "Memoir", p 137.

12 Galloway, "Memoir", p 138.

13 A.D. Jackson, "The Peaks Of The Middle Dart", *Apine Journal*, vol 6, no 22, Jun 1935, pp 36-7.

14 A. Dickie, "From The Hollyford To The Arawata", *Alpine Journal*, vol 5, no 21, Jun 1934, pp 308-9.

15 Swale, "Scripts".

16 Holloway, "Further Olivine Peaks And Passes", *Alpine Journal*, vol 7, no 25, Jun 1938, p 206.

17 I.C. McKellar, "A Month In The Olivine Country", *Outdoors*, Dunedin, Jul 1947, p 17.

18 A.J. Heine, "Journey Through O'Leary Pass", Alpine Journal, vol 16, no 42, Jun 1955, p 149.

19 Carter to Henderson.

20 J.D. Pascoe, *Great Days In New Zealand Exploration*, A.H. And A.W. Reed, Wellington, 1959, pp 172-3. See also Pascoe, *Explorers And Travellers, Early Expeditions In New Zealand*, A.H. And A.W. Reed, Wellington, 1983 (First Published As *Exploration New Zealand*, 1971), p vii. Pascoe, "Man Alone", In Henderson, "Home", *New Zealand Farmer*, vol 93, no 4, 24 Feb 1972, p 74. Pascoe, *The Haast Is In South Westland*, A.H. And A.W. Reed, Wellington, 1996, p 43. See also C. Maclean, *John Pascoe*, Craig Potton Publishing/The Whitcombe Press, Nelson, 2003, p 208.

21 James, "Journey". Sharpe, *Fiordland*, p 63.

22 Richardson, "Legendary".

23 Heine, "Arawata Adventure", *Alpine Journal*, vol 15, no 40, Jun 1953, p 214.

24 James, "5. A Search For Oil", *Otago Daily Times*, 8 Sep 1930, p 5, "Journey", "Mountains".

25 Holloway, "Plateau". Sealy, "Exploration", "Unknown".

26 Alexander, "Historical", pp 346-8. Dickie, "Hollyford", pp 308-9. Heine, "Arawata", p 214, "Journey", p 149. Holloway, "Bound", p 60-1, "Further", pp 206, 212. Jackson, "Peaks", pp 36-7. James, "Direct", p 132. Whitehead, "Ionia", pp 202, 207.

27 McCurdy, "Trip". Southland Tramping Club, *Jubilee*, p 22.

28 *Evening Star*, Undated Clipping, Author. See also *Otago Witness*, 22 Dec 1931, p 45.

29 Fortune, "Arawata", pp 289-91.

30 Chandler, "Arawata".

31 Gunn To Author.

32 His real name was Carl Axel Bjork but was variously anglicized as Karl Axel Borg, Yullus Berg, Jules Bjorg, Jules Borg and commonly Jules Berg.

33 Clark To Chandler, 14 Dec 1959, Chandler Papers, 4/6/10. Clark, "The Man Who Saw The Moas?", *Weekly News*, 21 Oct 1959, p 7.

34 Richardson, "Legendary".

35 Greaney, Heveldt, "Scripts". *Otago Daily Times*, 29 Jun 1946, p 8.

36 Fortune, "Passing", p 311. *Otago Daily Times*, 29 Jun 1946, p 8.

37 *Free Lance*, 5 Aug 1936. R.D. Climie To D.J.M. Glover, 30 Jun 1953, MS 0418-37, D.J.M. Glover, Papers, Alexander Turnbull Library, Wellington.

38 *Otago Daily Times*, 29 Jun 1946, p 8.

39 Gunn, "Arawata", p 183. *Free Lance*, 5 Aug 1936. R.D. Climie To D.J.M. Glover, 30 Jun 1953, MS 0418-37, D.J.M. Glover, Papers, Alexander Turnbull Library, Wellington.

40 McKenzie, *Pioneers*, p 78.

41 Swale, "Scripts".

42 F. Heffernan, "Scripts".

43 Cron, "Scripts".

44 Heveldt, "Scripts".

Chapter 10: Wakatipu Winters

1 Climie To Glover.

2 N.T. Cook To Author.

3 Cook To Author.

4 Cook To Author.

5 Cook To Author.

6 McDonald To Author.

7 McDonald To Author. M.E. Webb To Author.

8 McDonald To Author. Webb To Author.

9 McDonald To Author.

10 Dalley To Author. *Evening Post*, 25 Nov 1947, p 10. Koch, Martin, "Scripts".

11 B. Heffernan, Bourke, Martin, Swale, "Scripts".

12 McDonald To Author.

13 Koch, "Scripts".

14 F. Heffernan, "Scripts".

15 McEwan, Swale, "Scripts".

16 McDonald To Author. Webb To Author.

17 Speden To Author.

18 McDonald To Author. Webb To Author.

19 P. Bourne, *Out Of The World*, Geoffrey Bles, London, 1935, pp 236-8. Pamela Bourne refers to George (Gi) Shaw as Tom Shaw, probably confusing his first name with that of Tom Bryant further up the lake at Kinlock.

20 E. Bryant To Author. McKenzie, *Road*, pp 105-6. R. Bryant To Author.

21 Bourne, *Out*, p 237.

22 E. Bryant To Author. R. Bryant To Author.

23 Groves To Author.

24 R. Thomson To Author.

25 B. Heffernan, F. Heffernan, "Scripts".

26 Grant To Author.

27 Veint To Author. Webb To Author.

28 Sharpe To Author.

29 R.J. Meyer, *All Aboard, The Ships And Trains That Served Lake Wakatipu*, New Zealand Railway And Locomotive Society, Wellington, (First Edition 1963) Second Edition 1980.

30 R. Bryant To Author.

31 W.J. O'Leary To R. Moodie, Correspondence, Clark de Vries, Momona. Bob Moodie later reverted to his birth name of James McCarthy. When he died the prized letters were passed on to his grandson, Clark de Vries, at Momona near Dunedin. They are reproduced courtesy of Clark de Vries. Despite Bill O'Leary's hopes, Bob Moodie did lose the sight of an eye, the result of a bush felling accident near Glenorchy.

32 E. Bryant To Author. Groves To Author. McDonald To Author. R. Bryant To Author. Thompson To Author.

33 Forbes, Koch, "Scripts". Groves To Author. McDonald To Author. R. Bryant To Author. Thompson To Author.

34 Cron, "Scripts".

35 F. Heffernan, "Scripts".

36 McDonald To Author.

37 J. Hall-Jones, *Martins Bay*, Craig Printing, Invercargill, p 146. Richardson, "Legendary".

38 McDonald To Author. Speden To Author.

39 Sharpe To Author.

Chapter 11: Family And Flying

1 Thompson To Author.

2 Thompson To Author.

3 Thompson To Author.

4 Gunn To Author.

5 H. Wigley, *The Mount Cook Company, The First Fifty Years Of The Mount Cook Company*, Collins, Auckland, 1979, p 88.

6 Kennett, "Scripts". A. Munro To Gunn, In Gunn To Author.

7 Clingin, Forbes, Martin, "Scripts". Thompson To Author.

8 Richardson, "Legendary".

9 F. Heffernan, "Scripts".

10 McKenzie, *Road*, p 106.

11 Robb, *Hawea*, p 8.

12 Gunn To Author.

13 Speden To Author.

14 Richardson, "Legendary".

15 Munro To Gunn.

16 Fortune, "Passing", p 311. *Otago Daily Times*, 29 Jun 1946, p 8.

17 Cron, "Scripts". *Otago Daily Times*, 29 Jun 1946, p 8.

18 Cron, "Scripts". R. Bryant To Author.

19 *Otago Daily Times*, 29 Jun 1946, p 8.

20 Richardson, "Legendary".

21 McDonald To Author.

22 Dalley To Author. *Free Lance*, 19 Jun 1940, p 5.

23 Dalley To Author.

24 A. Owen, "Arawata Bill", In R.A. Knox (ed), *New Zealand's Heritage, The Making Of A Nation*, Paul Hamlyn, Wellington, 1971-3, vol 7 pt 91, pp 2540-1. A. Owen, "Arawata Bill", Spectrum

Documentary no 17, New Zealand Broadcasting Corporation, Wellington, 1972.

25 Dalley To Author.

26 *Free Lance*, 30 Aug 1939, p 5. O'Leary To Moodie, 1 Oct 1939, 15 Nov 1939, 20 Mar 1940.

27 Thompson To Author.

28 Dalley To Author.

29 Dalley To Author.

30 Dalley To Author.

31 Dalley To Author.

32 Dalley To Author.

33 O'Leary To Moodie, 20 Mar 1940.

34 Dalley To Author. *Otago Daily Times*, 29 Jun 1946, p 8.

35 Cron To Author. Robb, *Hawea*, p 8.

36 Coroner's Inquest Report, J46, 1900/763, National Archives Of New Zealand, Wellington. Death Records, Hokitika, 1900, no 45, Central Registry.

37 Coroner's Inquest Report, J46, 1907/756. Death Records, Wellington, 1907, no 588, Central Registry.

38 Dalley To Author.

39 Death Records, Hokitika, 1900, no 45, Central Registry. Timothy O'Leary's death records state that he had not five but four surviving sons and three daughters.

40 "Register Of Mining Rights, 1887-1900", Warden's Court, Queenstown, D98 Arc G11, National Archives, Dunedin, pp 20, 143, 152, 162, 185, 187, 251.

41 Dalley To Author.

42 F. Heffernan, Forbes, Kennett, "Scripts".

43 *Otago Daily Times*, 13 Apr 1893, p 3. See also 15 Apr 1893, p 2.

44 J. O'Leary To W.J.P. Hodgkins, 17 Mar 1894, MS 1164 R 30, New Zealand Alpine Club Archives, Hocken Library, Dunedin.

45 *Otago Daily Times*, 14 Mar 1894, p 2. This account was reprinted in the *New Zealand Alpine Journal*, vol 1, no 5, 1894, pp 297-8.

46 See for example W.S. Gilkison, *Earnslaw, Monarch Of Wakatipu*, Whitcombe And Tombs, Christchurch, 1957, pp 34-5.

47 Chandler, "Arawata". Chandler, *Head Of Lake Wakatipu Schools Centennial 1884-1984*, Centennial Committee, Glenorchy, 1984, p 36.

48 Dalley To Author.

49 Chandler, "Arawata". Dalley To Author. F. Heffernan, Kennett, "Scripts".

50 Dalley To Author.

Chapter 12: Glenorchy And Gold

1 Greaney, "Scripts".

2 Greaney, "Scripts".

3 A.D. Collins To Author.

4 Chandler, "Arawata". *Southland Tramper*, Invercargill, 1958, p 12.

5 Mrs I. Key To Gunn, 19 Mar 1972, Gunn Papers.

6 A.E. Paulin To Author. A.M. Yarker To Author. F. Yarker To Author. Kennett, "Scripts". R. Thomson To Author. T. Thomson To Author.

7   Shaw To Author. T. Thomson To Author. Veint To Author.

8   T. Thomson To Author.

9   A. Yarker To Author.

10 A. Yarker To Author. F. Yarker To Author. T. Thomson To Author.

11 Gunn To Author. "Mining Privileges Register, 1925-34", Hokitika Warden's Court, CH446
    15/7, National Archives Of New Zealand, Regional Office, Christchurch. R.J. Cuthill, "Bill's
    Wad Of Rights", *Weekly News*, 21 Oct 1959, p 7.

12 "Queenstown Register Of Miners Rights, 1933-(42)", Warden's Court, D98 JA 45/4, National
    Archives, Dunedin.

13 Gunn To Author.

14 Cook To Author. D. Nolan To Author. F. Heffernan, Martin, "Scripts". Fortune, "Passing", p
    311. F. Yarker To Author. Grant To Author. Gunn To Author. McKenzie, *Pioneers*, p 77. Nolan,
    Family, p 26. R. Bryant To Author. Sharpe, *Fiordland*, p 63. Speden To Author. T. Thomson To
    Author.

15 Cron To Author. Cron, "Scripts".

16 Fortune, "Passing", p 311. *Otago Daily Times*, 29 Jun 1946, p 8.

17 Bourke, "Scripts".

18 Martin, "Scripts".

19 *Free Lance*, 5 Aug 1936, p 53.

20 Richardson, "Legendary".

21 Cron, "Scripts".

22 *Free Lance*, 30 Aug 1939, p 5.

23 *Canterbury Mountaineer*, vol 4, no 17, 1948, p 240. Fortune, "Passing", p 310. *Free Lance*, 24
    Jul 1935, p 20, 30 Aug 1939, p 5. *Otago Daily Times*, 29 Jun 1946, p 8, 11 Nov 1947, p 6.

24 *Otago Daily Times*, 29 Jun 1946, p 8.

25 *Evening Star*, 31 Jul 1935, p 3.

26 *Free Lance*, 24 Jul 1935, p 20, 30 Aug 1939, p 5.

27 *Evening Star*, 10 Nov 1947, p 6. McDonald To Author.

28 Speden To Author.

29 Clingin, "Scripts".

30 J. Healy To Chandler, 6 Aug 1958, Chandler Papers, 4/6/10.

31 Cook To Author.

32 Koch, "Scripts".

Chapter 13: The Little Sisters

1   "Admission Book", Little Sisters Of The Poor Archives, Sacred Heart Home, Dunedin. F. Yarker
    To Author. McKenzie, *Pioneers*, p 78.

2   McKenzie, *Pioneers*, p 78.

3   G.M. Moir, "Pioneers Of The Hinterland", In *Evening Post*, 3 Dec 1947, p 12. *Otago Daily Times*,
    29 Jun 1946, p 8.

4   Fortune, "Passing", p 309. McKenzie, *Pioneers*, p 78. *Otago Daily Times*, 11 Nov 1947, p 6.

5   D. Watherston, "Scripts".

6   *Otago Daily Times*, 29 Jun 1946, p 8.

7   *Otago Daily Times*, 11 Nov 1947, p 6.

8   Bourke, Clingin, Cron, Elliot, Forbes, Kennett, "Scripts". R. Bryant To Author. Thompson To Author.

9   D. Nolan To Author. F. Heffernan, "Scripts". Grant To Author. Groves To Author. K. Nolan To Author. R. Thomson To Author. T. Thomson To Author.

10 B. Heffernan, "Scripts".

11 Beattie, *New Zealand*, p 38. Bill Beattie wrote of locating him at a "Salvation Army home in Dunedin."

12 Moir, "Hinterland".

13 Fortune, "Passing", p 309.

14 Fortune, "Passing", p 310.

15 Fortune, "Passing", p 312.

16 Greaney, "Scripts".

17 *Otago Daily Times*, 29 Jun 1946, p 8.

18 McKenzie, *Pioneers*, pp 77-8. See also pp 65-6.

19 *Otago Daily Times*, 29 Jun 1946, p 8.

20 Fortune, "Passing", p 309.

21 Kennett, "Scripts".

22 Dalley To Author.

23 Dalley To Author.

24 Dalley To Author.

25 Dalley To Author.

26 Dalley To Author.

27 Dalley To Author.

28 Greaney, "Scripts".

29 Fortune, "Passing", p 309. *Otago Daily Times*, 4 Jul 1947, p 6.

30 Collins To Author. Fortune, "Passing", p 309. *Free Lance*, 26 Nov 1947, p 47. *Otago Daily Times*, 4 Jul 1947, p 6, 11 Nov 1947, p 6. Speden To Author.

31 Collins To Author.

32 Collins To Author. Fortune, "Passing", p 309.

33 Greaney, "Scripts".

34 Speden To Author.

35 Duncan, "Arawhata".

36 Fortune, "Passing", p 309. *Otago Daily Times*, 4 Jul 1947, p 6.

37 Fortune, "Passing", p 309. *Otago Daily Times*, 4 Jul 1947, p 6.

38 Kennett, "Scripts".

39 Kennett, "Scripts".

40 Beattie, *Fiordland*, pp 84-5.

41 Kennett, "Scripts".

42 *Otago Daily Times*, 11 Nov 1947, p 6.

43 Deaths Records, Dunedin, 1947, no 952, Central Registry.

44 Cemetery Records, Andersons Bay Cemetery, Dunedin. Death Records, 1947, no 952, Central Registry. *Evening Star*, 10 Nov 1947, p 4. *Otago Daily Times*, 11 Nov 1947, p 1.

45 *Otago Daily Times*, 11 Nov 1947, p 6.

46 *Evening Star*, 10 Nov 1947, p 6.

47   *Southland Times*, 12 Nov 1947, p 4.
48   Richardson, "Legendary".
49   *Evening Post*, 25 Nov 1947, p 10.
50   T. Allen, "The Story Of 'Arawata Bill' ", In *Weekly News*, 19 Nov 1947, p 4.
51   *Free Lance*, 26 Nov 1947, p 47.
52   *Canterbury Mountaineer*, vol 4, no 17, 1948, p 240.
53   Fortune, "Passing", pp 309, 312.

Chapter 14: Plots And Plaques

1   Cemetery Records, Andersons Bay.
2   Chandler, "Arawata".   Chandler To T.A. Doogue, 23 Jul 1958, Chandler Papers, 4/6/10.
     Southland Tramper, 1958, p 12.
3   Chandler, "Arawata". *Southland Tramper*, 1958, p 12.
4   Sr Rosa To Chandler, 15 Apr 1958, Chandler Papers, 4/6/10.
5   "Admission Book", Little Sisters.
6   Chandler To *Southland Daily News* Editor, 9 May 1958, Chandler Papers, 4/6/10. *Southland
     Tramper*, 1958, p 12.
7   *New Zealand Truth*, Wellington, 15 Jul 1958, p 23. *Otago Daily Times*, 12 May 1958, p 1, 30
     May 1958, p 4.
8   Matheson To Chandler, 1 Jul 1958, Nolan To Chandler, Southland Tramping Club, "Subscription
     List For Memorials To 'Arawata Bill' & W.H. Homer", 4 Dec 1958, Stoop To Chandler, 19 Sep
     1958, Chandler Papers, 4/6/10.
9   Southland Tramping Club, "Subscription List", D. Scott To Chandler, 20 May 1958, E. Forsyth
     To Chandler, 2 Jul 1958, Dalley To Chandler, 1958, Chandler Papers, 4/6/10.
10   Chandler, "Arawata".
11   Chandler To Doogue.
12   *Otago Daily Times*, 30 May 1958, p 4.
13   A.E. Tilleyshort & Co Ltd To Chandler, 21 Aug 1958, Chandler Papers, 4/6/10.
14   Chandler To Lake County Clerk, 23 Apr 1959, Chandler Papers, 4/6/10.
15   Chandler To F. Heveldt, 23 Sep 1958, Chandler Papers, 4/6/10.
16   Chandler To Lake County.
17   Chandler To Lake County. Chandler (ed), *Moir's Guide Book, Northern Section*, New Zealand
     Alpine Club, Christchurch, Third Revised Edition 1961, p 27.
18   Chandler To Fiordland National Park Board Chairman, 29 Nov 1958, Chandler To *Truth*, 21
     May 1959, Chandler Papers, 4/6/10.
19   B. Heffernan, Elliot, F. Heffernan, "Scripts". D. Nolan To Author. Groves To Author. Sharpe To
     Author.
20   B. Heffernan, Elliot, F. Heffernan, "Scripts". Cook To Author. D. Nolan To Author. Groves To
     Author. Veint To Author.
21   Bourke, Clingin, Cron, F. Heffernan, Forbes, Koch, Watherston, "Scripts". Collins To Author.
     Cook To Author. Sharpe To Author. Shaw To Author. Thompson To Author. Grant To Author.
     R. Bryant To Author. Nolan, *Family*, pp 53, 64.
22   Bourke, Clingin, Cron, Forbes, Heveldt, Kennett, Martin, Watherston, "Scripts". McDonald
     To Author. K. Nolan To Author. R. Bryant To Author.

23 Jackson To Chandler, 16 Dec 1968, Chandler Papers, 4/6/10.

Chapter 15: Poetic Rhapsodies

1   D.J.M. Glover, *Arawata Bill, A Sequence Of Poems*, Pegasus Press, Christchurch, 1953.
2   Glover, *Hot Water Sailor 1912-1962 & Landlubber Ho 1963-1980*, Collins, Auckland, 1981 (*Hot Water Sailor* First Published By A.H. And A.W. Reed, Wellington, 1961), pp 77-8.
3   Glover, Pascoe, In Henderson, "Home", *New Zealand Farmer*, vol 93, no 4, 24 Feb 1972, p 74. Glover To T.J. Williams, 26 Feb 1953, MS 1418-121A, Glover, Papers. L. Cleveland, *The Horizon's Eye, Perspectives On The Poetry Of Denis Glover*, Kiwi Records, 1971. *New Zealand Listener*, Wellington, 17 Jul 1953, p 7.
4   Cleveland, *Horizon*. Glover, In Henderson, "Home". Glover To Williams, 26 Feb 1953. *Listener*, vol 29, no 731, 17 Jul 1953, p 7. In the later accounts, Denis Glover reverses the order in which the visits occurred. Denis Glover immediately sent John Pascoe and Rod Hewitt typed copies of the original poem. Rod Hewitt later gave his copy to his mountaineering partner Mavis Davidson, who pasted it inside a copy of *Arawata Bill, A Sequence Of Poems*, and in turn gave it to her friend Sheila Natusch, who kindly gave it to the author. A slightly altered version is also included in Denis Glover's papers in the Alexander Turnbull Library, MS 0418-121A, Glover, Papers.
5   Glover, In Henderson, "Home". Glover To Williams, 26 Feb 1953.
6   Glover To E.M. Burnett, 17 Sep 1973, E.M. Burnett Papers, MS 4420, Alexander Turnbull Library, Wellington.
7   Cleveland, *Horizon*. Glover, In Henderson, "Home". J.E.P. Thomson, *Denis Glover*, Oxford University Press, Wellington, 1977, p 8.
8   Glover To Williams, 26 Feb 1953.
9   Glover To Pascoe, 16 Feb 1953, MS 0418-121A, Glover, Papers.
10 Glover, *Arawata*, pp 11, 14, 36.
11 Glover, *Arawata*, pp 12-13, 16. Extracts from the poems are reproduced with the permission of the estate of the late Denis Glover and Richards Literary Agency.
12 Glover To Pascoe, 16 Feb 1953, Glover To Williams, 16 Feb 1953.
13 McKenzie, *Pioneers*. Notes Denis Glover took from the book are included in his papers in the Alexander Turnbull Library, MS 0418-121A.
14 Glover, *Arawata*, p 21. McKenzie, *Pioneers*, pp 77-8.
15 Glover, *Arawata*, p 19. McKenzie, *Pioneers*, p 78.
16 Glover, *Arawata*, pp 23, 25, 27-31, 37. McKenzie, *Pioneers*, p 77.
17 Glover, *Arawata*, p 26. *Free Lance*, 30 Aug 1939, p 5.
18 *Evening Post*, 25 Nov 1947, p 10. Glover, *Arawata*, pp 27-9.
19 Glover, *Arawata*, p 23. McKenzie, *Pioneers*, p 77.
20 Glover, *Arawata*, pp 12, 15, 23, 31, 33.
21 Glover, *Arawata*, pp 20, 23.
22 Glover To Pascoe.
23 *Listener*, vol 29, no 731, 17 Jul 1953, p 4.
24 Glover To Burnett.
25 Pascoe, *Haast*, p 43.
26 Pascoe, "Exploration, The Last Discoveries", In Knox, *Heritage*, vol 6, pt 83, 1971-3, p 2323. Pascoe, *Great*, p 146. Pascoe, *Haast*, p 43.
27 I. Agnew, *The Loner*, Hodder And Stoughton, Auckland, 1974.

28 Glover And G.W. Tricker, *Arawata Bill, Poems By Denis Glover, Etchings By Gary Tricker*, The Graphic Society Of New Zealand, Auckland, Limited Edition 1981.

29 Glover, *Enter Without Knocking, Selected Poems By Denis Glover*, Pegasus Press, Christchurch, (First Edition 1964) Enlarged Edition 1971, pp 83-102.

30 Williams To Glover, 12 Feb 1953.

31 *Listener*, vol 29, no 731, 17 Jul 1953.

32 Glover, *Arawata Bill And Other Verse*, Kiwi Records, 1971.

33 "Cover Notes", Glover, *Verse*.

34 Cleveland, *The Great New Zealand Songbook*, Godwit Press, Auckland, 1991, pp 36-8.

35 G. Ogilvie, *Denis Glover, His Life*, Random House, Auckland, 1999, p 426.

36 D.Q. Buchanan, Papers, FMS 4546-6, Alexander Turnbull Library, Wellington. D.Q. Buchanan To Author.

37 Pascoe, In Henderson, "Home".

38 *New Zealand Poetry Year Book*, Wellington, vol 3, 1953, pp 43-4. Pascoe, *Great*, pp 173-4. The poem is reproduced courtesy of John Pascoe's widow Dorothy Pascoe and Reed Publishing.

39 *Listener*, vol 29, no 731, 17 Jul 1953, p 7. Pascoe, *Great*, pp 173-4. *Poetry Year Book*, pp 43-4.

40 B. Ahern, "'Arawata Bill' (William O'Leary)", *Mount Aspiring National Park Newsletter*, Wanaka, no 3, Sep 1982, p 10. The peom is reproduced courtesy of Brian Ahern.

41 Ahern, "Arawata", p 14

42 *Otago Daily Times*, 17 May 1958, p 4.

43 *Truth*, 15 Jul 1958, p 23.

Chapter 16: Paddocks And Paintings

1 Sharpe, *Fiordland*, p 63.

2 Sharpe, *Fiordland*, pp 63-4.

3 Henderson, "Home", In *New Zealand Farmer*, vol 93, no 1, 13 Jan 1972, p 55, no 4, 24 Feb 1972, pp 74-5, no 6, 23 Mar 1972, p 69, no 9, 11 May 1972, p 72.

4 *Listener*, vol 29, no 731, 3 Jul 1972, p 36. Owen, *Spectrum*.

5 Owen, "Arawata Bill", In Knox, *Heritage*, vol 7, pt 91, pp 2538-41. Owen, "O'Leary, William", In C. Orange (ed), *Dictionary Of New Zealand Biography, Volume Three, 1901-1920*, Auckland University Press/Department Of Internal Affairs, Wellington, 1996.

6 Thompson To Author.

7 Thompson to Author.

8 Veint To Author

9 *Stone's*, 1954, p 935.

10 McKenzie, *Road*, p 106.

11 *Evening Star*, 10 Nov 1947, p 4. *Otago Daily Times*, 11 Nov 1947, p 1.

12 Stoop To Chandler, 15 Sep 1958.

13 Fortune, "Passing", p 309.

14 McKenzie, *Pioneers*, pp 38-41.

15 McKenzie, *Pioneers*, pp 38-41.

16 McKenzie, *Pioneers*, pp 38-41.

17 McKenzie, *Pioneers*, pp 38-41. Pickering, *Southern*, p 85.

18 Webster, "Diary", 13 Nov 1912.

19 McKenzie, *Pioneers*, Maori Bill, pp 38-41, Arawata Bill, pp 65-66, 77-8.

20 Pascoe, *Mr Explorer Douglas*, A.H. And A.W. Reed, Wellington, 1957, p 79.

21 I.O. Roxburgh, *Jacksons Bay, A Centennial History*, A.H. And A.W. Reed, Wellington, 1976, pp 84-5.

22 N. Peat, *Cascade On The Run*, Whitcoulls, Christchurch, 1979, p 68.

Chapter 17: Everlasting Whisper

1 Fortune, "Arawata", p 290.

2 *Evening Post*, 25 Nov 1947, p 10.

3 Fortune, "Passing", pp 311-12.

4 Greaney, "Scripts".

BIBLIOGRAPHY

Official Records

Appendix To The Journals Of The House Of Representatives, Wellington.

Births, Deaths And Marriages Records, Births, Deaths and Marriages Central Registry, Lower Hutt.

Births, Deaths And Marriages Records, District Court, Balclutha.

Cemetery Records, Andersons Bay Cemetery, Dunedin.

Coroners' Inquest Reports, National Archives Of New Zealand, Wellington.

Electoral Rolls, House Of Representatives.

Lawrence Area School Papers, Hocken Library, Dunedin.

List Of Immigrants, Debtors To The Provincial Government Of Otago For Passage Moneys, 1869.

Maps, Department Of Lands And Survey (Terralink), Wellington.

New Zealand Geographic Board Records, New Zealand Geographic Board, Wellington.

Passenger Lists Victoria Australia Onwards To New Zealand Ports, Otago Settlers Museum, Dunedin.

Wardens Courts' Records. National Archives Of New Zealand, Regional Office, Christchurch.

Wardens Courts' Records. National Archives Of New Zealand, Regional Office, Dunedin.

Unofficial Records and Private Papers

Buchanan, Dorothy Quita, Papers, Alexander Turnbull Library, Wellington.

Burnett, Ethelwyn Moana, Papers, Alexander Turnbull Library, Wellington.

Catholic Church Otago-Southland Archives, St Joseph's Cathedral Presbytery, Dunedin.

Chandler, Peter Mitchell, Papers, Hocken Library, Dunedin.

Douglas, Charles Edward, Papers, Alexander Turnbull Library, Wellington.

Glover, Denis James Matthews, Papers, Alexander Turnbull Library, Wellington.

Gunn, Murray Alexander, Papers, Hollyford Museum, Hollyford.

Little Sisters Of The Poor Archives, Sacred Heart Home, Dunedin.

McCurdy, Lindsay Frederick, "Invercargill Trip Book", Margaret Ann Irving (Ann), Waimatua.

McKenzie, Daniel, "1903 Diary", Ian Gordon Speden, Lower Hutt.

Natusch, Sheila Ellen, Papers, Alexander Turnbull Library, Wellington.

New Zealand Alpine Club Papers, Hocken Library, Dunedin.

O'Leary, William James (Bill) To Moodie, Robert (Bob), Correspondence, Clark de Vries, Momona.

Owen, Alwyn (Hop), "Scripts For Arawata Bill Broadcast On NZBC", Alexander Turnbull Library, Wellington.

Turner, Francis John (Frank), "Geological Expeditions To Haast-Cascade Areas, 1929, 1930", Alan Frederick Cooper, Dunedin.

Webster, James Deans, "Diary", 1912, Hocken Library, Dunedin.

Newspapers, Periodicals and Annuals

Canterbury Mountaineer, Christchurch.

Christchurch Star-Sun, Christchurch.

Evening Post, Wellington.

Evening Star, Dunedin.

Hokitika Guardian, Hokitika.

Lake Wakatip Mail Queenstown.

Massif, Palmerston North.

Mataura Ensign, Gore.

Mount Aspiring National Park Newsletter, Wanaka.

New Zealand Alpine Journal, Dunedin.

New Zealand Farmer, Auckland.

New Zealand Free Lance, Wellington.

New Zealand Geographic, Ohakune.

New Zealand Listener, Wellington.

New Zealand Poetry Year Book, Wellington.

New Zealand Railways Magazine, Wellington.

New Zealand Truth, Wellington.

(New Zealand) Weekly News, Auckland.

Otago Colonist, Dunedin.

Otago Daily Times, Dunedin.

Otago Witness, Dunedin.

Outdoors, Dunedin.

Southland Daily News, Invercargill.

Southland Times, Invercargill.

Southland Tramper, Invercargill.

Stone's Otago And Southland Directory, Dunedin.

Tuapeka Times, Lawrence.

Wide World Magazine, London.

Books and Pamphlets

Agnew, Ivan, *The Loner,* Hodder And Stoughton, Auckland, 1974.

Beattie, James Herries (Herries), *Far Famed Fiordland,* Otago Daily Times And Witness Newspapers Co, Dunedin, 1950.

Beattie, William B. (Bill), *Bill Beattie's New Zealand,* Hodder And Stoughton, Auckland, 1970.

Bishop, Catherine Jane (Jane) And Walker, Malcolm, *Westland County, A Centennial Album 1876-1976,* Westland County Council, 1977.

Bourne, Pamela, *Out Of The* World, Geoffrey Bles, London, 1935.

Bradshaw, Julia, *Miners In The Clouds, A Hundred Years Of Scheelite Mining At Glenorchy,* Lakes District Museum, Arrowtown, 1997.

Bradshaw, Julia, *The Far Downers, The People And History Of Haast And Jackson Bay,* University of Otago Press, Dunedin, 2001.

Catholic Church, *The Lode Of Faith, The Church In The Tuapeka, St Patricks Lawrence, Centennial Souvenir, A Diocesan Centennial Publication,* Lawrence, 1972.

Chandler, Peter Mitchell, *Head Of Lake Wakatipu Schools Centennial 1884-1984,* Centennial Committee, Glenorchy, 1984.

Chandler, Peter Mitchell, *Land Of The Mountain And The Flood, A Contribution To The History Of Runs And Runholders Of The Wakatipu District,* Queenstown And District Historical Society, Queenstown, 1996.

Chandler, Peter Mitchell (ed), *Moir's Guide Book, Northern Section,* New Zealand Alpine Club, Christchurch, Third Revised Edition 1961 (See also Chandler And Keen, Gilkison And Hamilton).

Chandler, Peter Mitchell, And Keen, Ronald J. (Ron) (eds), *Moir's Guide Book, Northern Section,* New Zealand Alpine Club, Christchurch, Fourth Revised Edition 1968.

Cleveland, Leslie (Les), *The Great New Zealand Songbook,* Godwit Press, Auckland, 1991.

Cleveland, Leslie (Les), *The Horizon's Eye, Perspectives On The Poetry Of Denis Glover,* Kiwi Records, 1971.

Ell, Gordon, *Gold Rush, Tales And Traditions Of The New Zealand Goldfields,* Bush Press, Auckland, 1995.

Ell, Gordon, *New Zealand Traditions And Folklore,* Bush Press, Auckland, 1994.

Fulton, Myra, *Okura: The Place Of No Return,* Myra Fulton, Takaka, 2004.

Gilkison, Walter Scott (Scott), And Hamilton, Arthur Hunter (eds), *Moir's Guide Book,* New Zealand Alpine Club, Dunedin, Second Revised And Enlarged Edition 1948 (First Published As Moir, George Morrison, *Guide Book To The Tourist Routes Of The Great Southern Lakes,* Otago Daily Times And Witness Newspapers Co, Dunedin, 1925).

Gilkison, Walter Scott (Scott), *Earnslaw, Monarch Of Wakatipu,* Whitcombe And Tombs, Christchurch, 1957.

Gilkison, Walter Scott (Scott) (ed), *Handbook To The Mount Aspiring National Park,* Mount Aspiring National Park Board, Dunedin, 1971.

Glover, Denis James Matthew, *Arawata Bill, A Sequence Of Poems,* Pegasus Press, Christchurch, 1953.

Glover, Denis James Matthew, *Enter Without Knocking, Selected Poems By Denis Glover,* Pegasus Press, Christchurch, (First Published 1964) Enlarged Edition 1971.

Glover, Denis James Matthew, *Hot Water Sailor 1912-1962 And Landlubber Ho* 1963-1980, Collins, Auckland, *1981 (Hot Water Sailor* First Published By A.H. And A.W Reed, Wellington, 1961).

Glover, Denis James Matthew, And Tricker, Gary Walter, *Arawata Bill, Poems By Denis Glover, Etchings By Gary Tricker,* The Graphic Society Of New Zealand, Auckland, Limited Edition 1981.

Gordon, John And Rundle, John, *Mountains Of The South,* Random House New Zealand, Auckland, 1993.

Hall-Jones, John, *Fiordland Placenames,* Fiordland National Park Board, Invercargill, 1979.

Hall-Jones, John, *Martins Bay,* Craig Printing, Invercargill, 1987.

Henderson, James Herbert (Jim) (ed), *Open Country Muster, People And Places Out Of Town,* A.H. And A.W Reed, Wellington, 1974.

Jenkin, Robyn, *New Zealand Mysteries,* A.H. And A.W Reed, Wellington, 1970.

Johnston, Gordon Wills, *Mount Aspiring National* Park, Bascands, Christchurch, 1976.

Kay, Rupert, *Westland's Golden Century 1860-1960, An Official Souvenir Of Westland's Centenary,* Westland Centennial Council, Greymouth, 1960.

Knox, Raymond Anthony (Ray) (ed), *New Zealand's Heritage, The Making Of A Nation,* Paul Hamlyn, Wellington, 1971-1973.

Mayhew, William Richardson, *Tuapeka, The Land And Its People, A Social History Of The Borough Of*

Lawrence And Its Surrounding Districts, Otago Centennial Historical Committee, Dunedin, 1949.

Maclean, Chris, *John Pascoe*, Craig Potton Publishing/The Whitcombe Press, Nelson, 2003.

McGill, David Keith, *Ghost Towns Of New Zealand*, A.H. And A.W. Reed, Wellington, 1980.

McKenzie, Alice (Mrs Peter Mackenzie), *Pioneers Of Martins Bay*, Southland Historical Committee, Invercargill, 1947.

McKenzie, Doreen, *Road To Routeburn*, John McIndoe, Dunedin, 1973.

McLauchlan, Gordon (ed), *The Illustrated Encyclopedia Of New Zealand*, David Bateman, Auckland, 1989.

Meyer, Robert John, *All Aboard, The Ships And Trains That Served Lake Wakatipu*, New Zealand Railway And Locomotive Society, Wellington, (First Published 1963) Second Edition 1980.

Nolan, James Patrick, *A Saga Of The South, A Short History Of A West Coast Family*, The Nolans, Nolan Reunion Committee, Christchurch, 1983.

Nolan, Maxwell Albert (Tony), *Gold Trails Of The West Coast*, A.H. And A.W. Reed, Wellington, 1975.

Nolan, William Denis (Bill), *The Nolan Family, Pioneers Of South Westland 1871-1993*, Thornton Publishing, Christchurch, 1993.

Ogilvie, Gordon, *Denis Glover, His Life*, Random House, Auckland, 1999.

Ogilvie, Gordon, *Introducing Denis Glover*, Longman Paul, Auckland, 1983.

Orange, Claudia (ed), *Dictionary of New Zealand Biography, Volume Three, 1901-1920*, Auckland University Press/Department of Internal Affairs, Wellington, 1996, *Volume Four, 1921-1940*, 1998.

Pascoe, John Dobree, *Explorers And Travellers, Early Expeditions In New Zealand*, A.H. And A.W. Reed, Wellington, 1983 (First Published As *Exploration New Zealand*, 1971).

Pascoe, John Dobree, *Great Days In New Zealand Exploration*, A.H. And A.W. Reed, Wellington,1959.

Pascoe, John Dobree, *The Haast Is In South Westland*, A.H. And A.W. Reed, Wellington, 1966.

Pascoe, John Dobree (ed), *Mr Explorer Douglas*, A.H. And A.W. Reed, Wellington, 1957.

Peat, Neville, *Cascade On The Run*, Whitcoulls, Christchurch, 1979.

Peat, Neville, *Land Aspiring, The Story of Mount Aspiring National Park*, Craig Potton, Nelson, 1994.

Pickering, Mark, *The Southern Journey*, Mark Pickering, Christchurch, 1993.

Robb, Eileen, *Hawea Hospitality*, Eileen Robb, Wanaka, 1986.

Roxburgh, Irvine Owen, *Jacksons Bay, A Centennial History*, A.H. And A.W. Reed, Wellington, 1976.

Salmon, John Hearsey McMillan, *A History Of Goldmining In New Zealand*, Government Printer, Wellington, 1963.

Shadbolt, Maurice And Brake, Brian, *Reader's Digest Guide To New Zealand*, Reader's Digest, Sydney, 1988.

Shadbolt, Maurice (ed), *The Shell Guide To New Zealand*, Whitcombe And Tombs, Christchurch, 1968.

Sharpe, Rupert James Drummond, *Fiordland Muster*, Hodder And Stoughton, London, 1966.

Southland Tramping Club, *Southland Tramping Club, 25th Jubilee, Southland Tramping Club Inc 1947-1972*, Southland Tramping Club, Invercargill, 1972.

Spearpoint, Geoff, *Waking To The Hills, Tramping In New Zealand*, Reed Methuen, Auckland, 1985.

Thomson, Jane (ed), *Southern People, A Dictionary Of Otago Southland Biography*, Longacre Press/ Dunedin City Council, Dunedin, 1998.

Thomson, John Edward Palmer, *Denis Glover*, Oxford University Press, Wellington, 1977.

Wigley, Henry (Harry), *The Mount Cook Way, The First Fifty Years Of The Mount Cook Company*, Collins, Auckland, 1979.

Interviews and Correspondence

Anderson, Andrew (Andy), Christchurch.

Bryant, Ernest Valpy (Ernie), Alexandra.

Bryant, Richard Henry (Harry), Frankton.

Buchanan, Dorothy Quita, Wellington.

Buchanan, Henry John, Waiatoto.

Collins, Albert David (Albie), Albert Town.

Cook, Neil Tasman, Mosgiel.

Cron, Allen Andrew, Blenheim.

Dalley, Stella Mary (Stella Somes), Wellington.

Eggeling, Milcah Elizabeth (Betty Buchanan), Okuru.

Gilkison, James Hogg (Jim), Dunedin.

Grant, Rose Frances (Rosie), Glenorchy.

Groves, Norrie Ivan, Warrington.

Gunn, Murray Alexander, Hollyford.

Jackson, Brian Andrew, St Bathans.

McDonald, June Laura (Elfin Shaw), Dipton.

Nolan, Desmond Joseph (Des), Okuru.

Nolan, Kevin, Okuru.

Nolan, William Denis (Bill), Christchurch.

Paulin, Andrew Elliot (Andy), Rangiora.

Sharpe, Peter Gerald, Queenstown.

Shaw, Mark Frederick, Gillespies Beach.

Speden, Gordon, Gore.

Thompson, Robert Curwen (Bob), Queenstown.

Thomson, Reta (Reta Groves), Glenorchy.

Thomson, Thomas James (Tommy), Glenorchy.

Veint, Charles Lloyd (Lloyd), Queenstown.

Webb, Moira Ellen (Moira Crowe), Invercargill.

Yarker, Ailsa Maud, Hampden.

Yarker, Francis Victor (Vic), Hampden.

Audio Recordings

Glover, Denis James Matthew, *Arawata Bill And Other Verse*, Kiwi Records, 1971.

Greaney, Dan, Heveldt, Frank, Recorded Interviews, New Zealand Public Radio Sound Archives, Christchurch.

Owen, Alwyn (Hop), *Arawata Bill*, Spectrum Documentary No 17, New Zealand Broadcasting Corporation, Wellington, 1972.

Illustrations

Ahern, Brian Richard, Wanaka.

Alexander Turnbull Library, Wellington.

Balclutha District Court, Balclutha.

Bishop, David Graham (Graham), Dunedin.

Canterbury Museum, Christchurch.

Carr, Clare Joanne, Queenstown.

Christchurch Star-Sun, Christchurch.

Dalley, Stella Mary, Wellington.

de Vries, Clark, Momona.

Department Of Conservation, Wanaka.

Dougherty, Ian Murray, Dunedin.

Entwistle, Robert Field (Bob), Dunedin.

Gunn, Murray Alexander, Hollyford.

Hocken Library, Dunedin.

Kelk, John, Invercargill.

Looking At The West Coast, Greymouth.

Mason, Kenneth David (Ken), Dunedin.

National Archives Of New Zealand, Regional Office, Dunedin.

Otago Settlers Museum, Dunedin.

Otago Witness, Dunedin.

Out Of The World, London.

Southland Daily News, Invercargill.

Speden, Ian Gordon, Lower Hutt.

Thompson, Robert Curwen, Queenstown.

Tricker, Gary Walter, Greytown.

Webb, Moira Ellen, Invercargill.

Weekly Press, Christchurch.

Wide World Magazine, London.

INDEX

(Opposite) The original manuscript of the poem Arawata Bill which Denis Glover gave to Rod Hewitt, who shaved in the mountains. (I.M. Dougherty.)